FrameMaker v18.0.0 Structured EDD Development

A workbook for self-paced or instructor-led training

Matt R. Sullivan

ISBN: 978-1-953488-03-9

Important Notice

Written and designed at Adobe Systems Incorporated, 345 Park Avenue, San Jose, CA 95110 94039-7900, USA

Adobe Systems Europe Limited, Adobe House, 5 Mid New Cultins, Edinburgh EH11 4DU, Scotland, United Kingdom

Adobe Systems Co., Ltd., Yebisu Garden Place Tower, 4-20-3 Ebisu, Shibuya-ku, Tokyo 150, Japan

About the Author

Matt Sullivan is the founder of *Tech Comm Tools*. Along with the FrameMaker training and consulting available at www.techcommtools.com, Matt helps organizations deliver content (including video and interactive media within their documentation) to online and mobile users.

Matt is an Adobe Tech Comm Partner, Adobe Certified Trainer, Adobe Certified Expert, and an Adobe Community Professional. He has produced (for Adobe) new feature videos for many versions of FrameMaker and the Adobe Technical Communication Suite.

He is also the author of

- FrameMaker - Structured Authoring Workbook (v18/2026 Release)
- FrameMaker - Structured Authoring Workbook (v16/2020 Release)
- FrameMaker - Structured Authoring Workbook (v15/2019 Release)
- FrameMaker - Structured Authoring Workbook (v14/2017 Release)
- FrameMaker - Structured EDD Development Workbook (v18, this book, released in 2026)
- FrameMaker - Structured EDD Development Workbook (v16/2020 Release)
- FrameMaker - Structured EDD Development Workbook (v15/2019 Release)
- FrameMaker - Structured EDD Development Workbook (v14/2017 Release)
- FrameMaker - Working with Content reference book (v16/2020 Release)
- FrameMaker - Working with Content reference book (v14/2017 Release)
- FrameMaker - Creating and Editing Content reference book (v13/2015 Release)
- Publishing Fundamentals: Unstructured FrameMaker reference book (v11/2012 Release)

You can find information about print and eBooks of these titles at
https://learn.techcommtools.com/collections/books

Both structured and standard (unstructured) courses are available at
https://courses.techcommtools.com/

Please connect with Matt via *mattrsullivan* on social platforms.

When not working with clients, you'll find Matt fixing things around the house, and refereeing college and semi-pro soccer.

Reach Matt directly by emailing him at matt@techcommtools.com

Stay up-to-date with Matt's Adobe books, webinars, and videos by signing up for his free newsletter at
https://techcommtools.com/email-list

Contents

Chapter 3: GeneralRule for Containers and Footnotes

Chapter 4: GeneralRule for Tables and Table Parts

Chapter 5: Tables—InitialStructurePattern and InitialTableFormat

Chapter 6: Inclusions and Exclusions

Chapter 7: AutoInsertions

Chapter 8: Defining and Formatting Objects

Chapter 9: Attribute List

Chapter 10: Formatting with AllContextsRules

Chapter 11: ContextRule formatting rules

Chapter 12: First/LastParagraphRules

Chapter 13: PrefixRules and SuffixRules

Chapter 14: Elements for Structuring Books

Chapter 15: Structuring Unstructured Data

Chapter 1: Getting started

Introduction

In this module, you will learn how FrameMaker works with structured content models, identify how to get started creating an Element Definition Document, identify the various types of elements you can define and the various parts of an element definition.

Objectives

- Download class files referenced in exercises
- Define structured documentation
- Review overall development process
- Identify several ways to create initial EDD
- Identify basic types of elements you can define
- Identify parts of element definition

Downloading class files

The first few lessons utilize sample files, and later lessons have incremental files available for your use to help keep you aligned with the workbook.

If class files haven't already been downloaded for your use, you'll want to download them before starting the lessons. Use the QR code shown to download the file or complete Exercise 1 before continuing.

Exercise 1: Downloading class files

To download the lesson files:

1. Open a browser like Chrome or FireFox and navigate to https://techcomm.tools/edd-dev
2. Provide an email address to get the files sent to you via email.

   A confirmation message will be sent to ensure you've entered the proper address.
3. Answer the confirmation message in the affirmative to receive the file download link for your file.
4. Decompress the file sent via email to your preferred location.

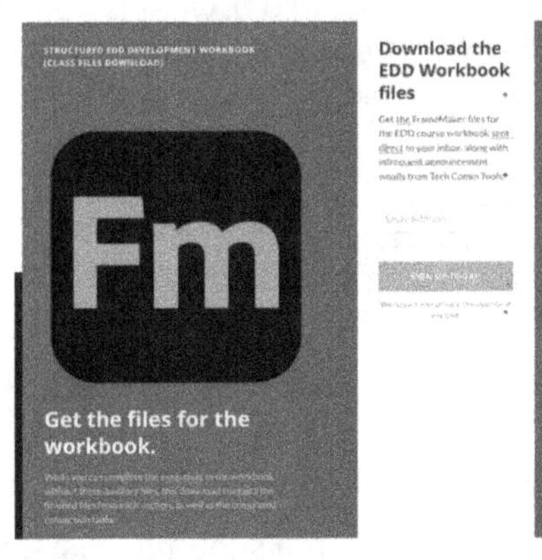

Choose a location that is easy for you to remember and to type (like the root of the C: drive, your desktop, or your Documents folder) as you'll need to navigate to these files throughout the course.

If you are using FrameMaker version 18.0.0 or later you should use the version 18 files. Earlier versions are also provided, if you are not yet using FrameMaker 18.0.0 or later.

Check **Help > About FrameMaker** to confirm you currently installed version of FrameMaker.

Understanding how an EDD controls structure and formatting

Structured documents are organized into logical chunks of information, called elements. Elements have rules related to their:

- Content
- Hierarchy
- Order
- Frequency
- Necessity (Required or Optional)

A content model in XML and SGML environments is defined by either a Document Type Definition (DTD) or a Schema, with other technologies controlling formatting.

FrameMaker uses a document called an Element Definition Document (EDD) to manage both the content model and the formatting of the content itself.

Authors focus on content and organization of information, not formatting. Focusing on content promotes documentation consistency throughout companies and across industries.

Consistency improves communication of information.

In FrameMaker, authors create and edit documents using elements whose definitions are:

- Defined in a corresponding EDD
- Imported into a design template, populating an element catalog in the design template. The element catalog, along with other traditional FrameMaker components like master pages and other formatting information make up a structured template.

As of 18.0.0, what is commonly refered to as the *Element Catalog* is displayed in the *Elements* panel. These phrases are used interchangeably in this book, at least until Adobe changes the label of either the menu item or the *Elements* panel.

Creating an Initial EDD

There are several ways to create an initial EDD:

- If you are conforming to an existing DTD, you can convert a DTD to an EDD.
- If no DTD exists, but you have a structured FrameMaker document containing an element catalog, you can export the element catalog as an EDD from your structured document.
- If a schema exists, you can import the schema into FrameMaker as an EDD.
- If no DTD and no comparable structured document, you can start with a new, empty EDD.

After a brief review of DTD and EDD files, you'll start in this workbook with a new, empty EDD.

Exercise 2: Exploring a raw DTD

For instructions on downloading class files, see "Downloading class files" on page 1.

In this exercise, you will open a sample DTD, without converting it to an EDD, and view it as a text file.

1. In FrameMaker, from your **EDDClassFiles\EDD-WB-Files-2026-18-0-0a** directory, open **xml-dtd.dtd.**
   a. From the **File** menu, choose **Open.**

      The **Open** dialog appears.

      If necessary, change to your class files directory.

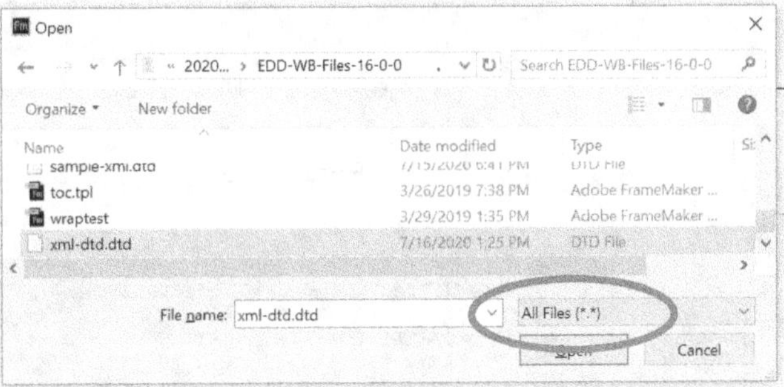

If necessary, choose **All Files (*.*)** from the filetype menu.

   b. Double-click **xml-dtd.dtd.**

      The **Unknown File Type** dialog appears.

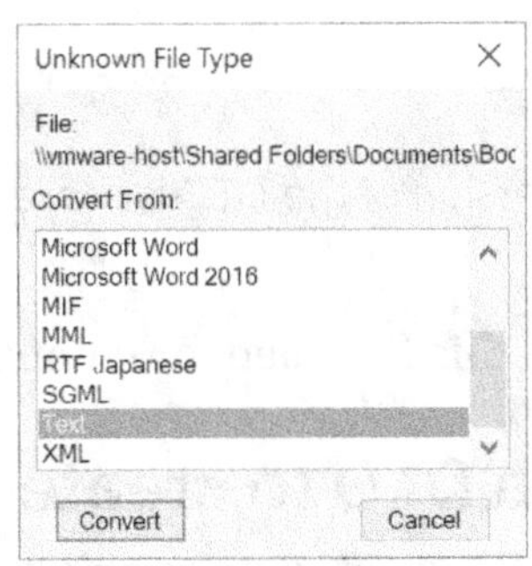

   c. Select **Text** and click **Convert.**

      The **Reading Text File** dialog appears.

   d. Select **Treat Each Line as a Paragraph** and click **Read.**
      The sample DTD appears.

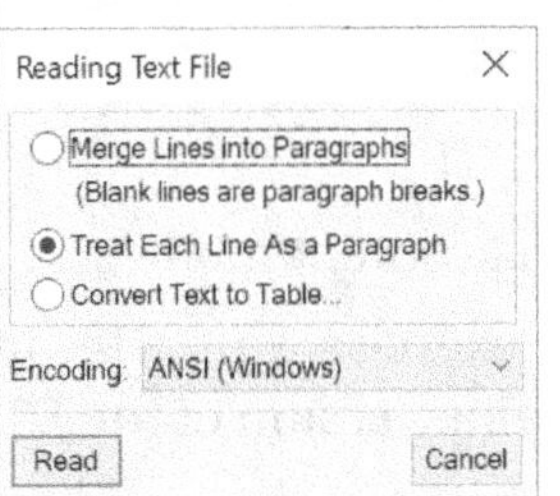

2. Scroll through the DTD, reviewing its contents.

```
<!--DTD for Chapter. Typically invoked by
      <!DOCTYPE  Chapter  SYSTEM "c:\class\sample-xml.dtd">
-->

<!--Chapter: Chapter is container of all elements within individual chapter of
 maintenance manual-->
<!ELEMENT Chapter  (Title, Section, Section+) >
<!ATTLIST Chapter    Author     CDATA     #REQUIRED
                     Version    NMTOKEN   "1.0" >
<!ELEMENT Title   (#PCDATA) >
<!ATTLIST Title  ID ID  #IMPLIED >
<!ELEMENT Section  (Head, (((Para, List?)+, (Section, Section+)?) |
                          (Section, Section+))) >
<!ELEMENT Head      (#PCDATA) >
<!ATTLIST Head       ID        ID        #IMPLIED >
<!ELEMENT List      (Item, Item+) >
<!ATTLIST List      ListType   (Bulleted|Numbered)   "Bulleted" >
<!ELEMENT Item       (Para+) >
<!ELEMENT Para       (#PCDATA) >
```

The text version of the DTD is more like application code, and not terribly human-friendly.

3. Leave this file open for comparison with the file in the next exercise.

Processing a DTD to create an EDD

FrameMaker gives you the ability to process the following into unformatted EDDs:

- XML DTDs
- SGML DTDs
- Schema content models

In the next exercise, you will convert the DTD from the last chapter into an EDD, which provides a more readable version of the DTD code.

Exercise 3: Converting a DTD into an EDD

In this exercise, you will open a sample DTD, which converts it to an EDD, and view the resulting element definitions.

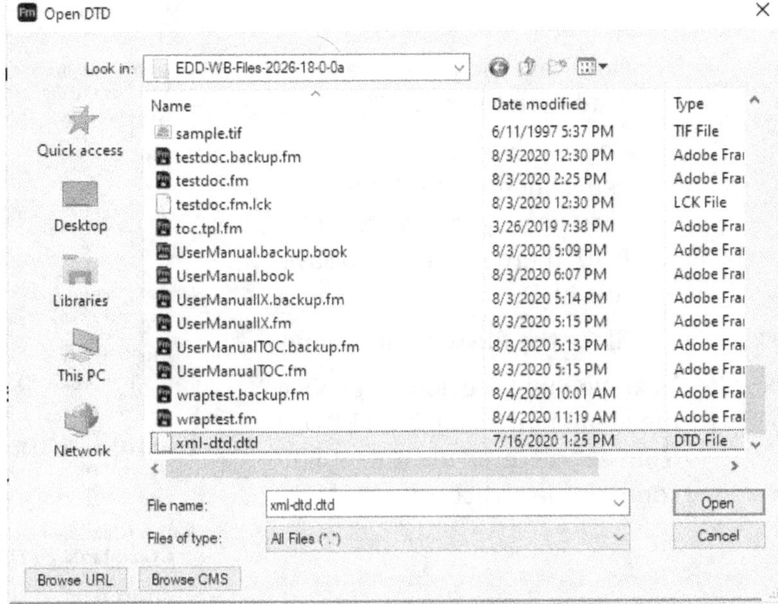

1. From the menu bar, select **Structure>DTD > Open DTD**.

   The **Open DTD** dialog appears.

2. If necessary, change to your class files directory.

3. Double-click **xml-dtd.dtd**.

   The **Use Structured Application** dialog appears.

4. Choose **<No Application>** from the **Use Structured Application** popup menu, and click **Continue**.

   The **Select Type** dialog appears.

5. Select **XML** (if necessary) and select **OK**.

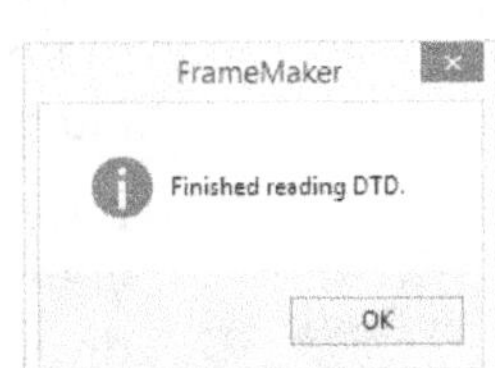

   An alert appears with the message, “Finished reading DTD.”

6. Click **OK** to dismiss alert.

   An untitled EDD appears with element and attribute definitions corresponding to original DTD.

7. Compare this EDD with the DTD you explored in Exercise 2.

   Note that the EDD is easier to read than the raw DTD.

   Because, you are experimenting with the various ways to create an EDD, there's no need to save this EDD.

8. Close the file without saving.

Next, you will see how to export the element catalog as an EDD from a structured document with comparable structure.

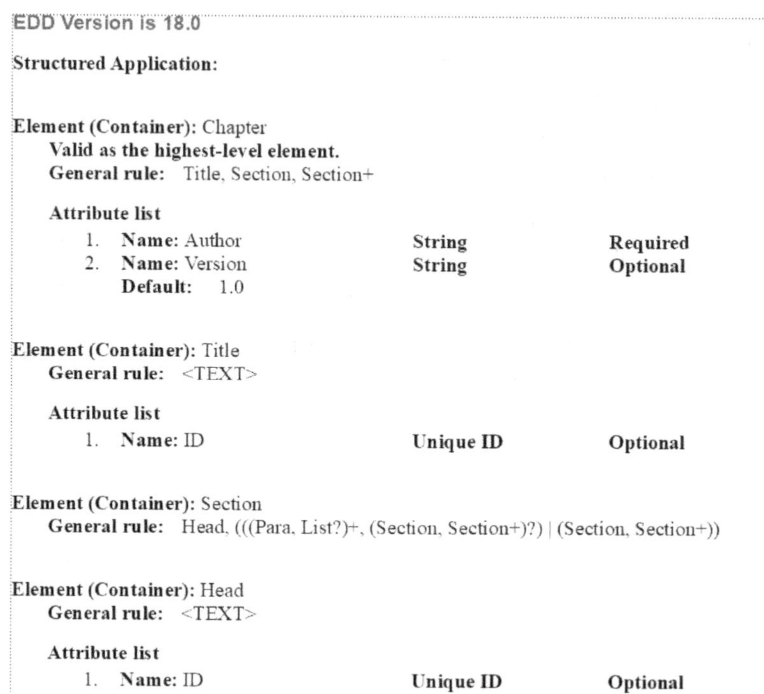

EDD Version is 18.0

Structured Application:

Element (Container): Chapter
Valid as the highest-level element.
General rule: Title, Section, Section+

Attribute list

1.	**Name:** Author	**String**	**Required**
2.	**Name:** Version **Default:** 1.0	**String**	**Optional**

Element (Container): Title
General rule: <TEXT>

Attribute list

1.	**Name:** ID	**Unique ID**	**Optional**

Element (Container): Section
General rule: Head, (((Para, List?)+, (Section, Section+)?) | (Section, Section+))

Element (Container): Head
General rule: <TEXT>

Attribute list

1.	**Name:** ID	**Unique ID**	**Optional**

Exercise 4: Exporting an EDD from a structured FrameMaker Document

In this exercise, you will export the element catalog as an EDD from a structured document.

1. From your class files directory, open **fm-sample.fm**.

   a. Display the **Open** dialog by choosing **File>Open**.

   b. If necessary, change to your class files directory.

   c. Double-click **fm-sample.fm**. Dismiss any missing resource dialogs.

   The sample structured document appears.

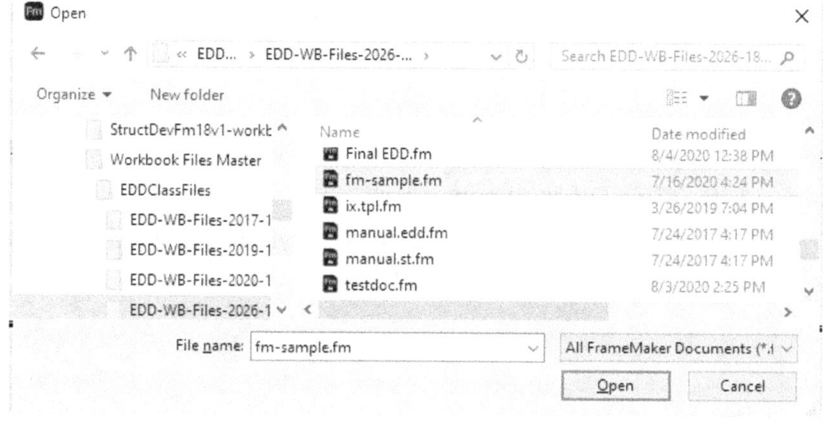

2. Scroll through the sample document, viewing its contents.

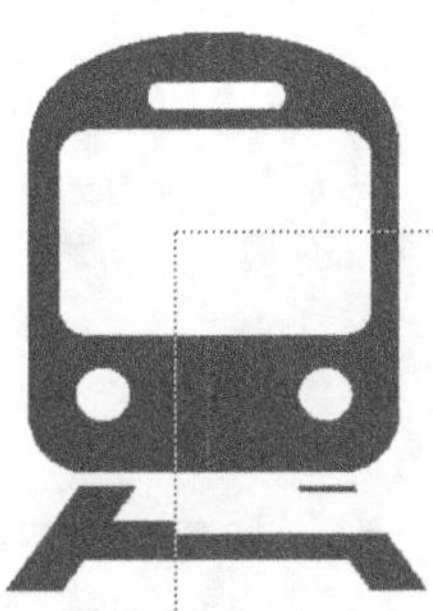

Chapter 9
Side Doors

Joe Smith
Technical Documentation
April 15, 2026

9.1 INTRODUCTION

9.1.1 Chapter Overview

9.1.1.1 Procedures in This Chapter

This chapter describes maintenance procedures for the side doors on the AstroLiner T440B and T442 light rail cars. It includes safety guidelines, an overview of door components, and a maintenance schedule for some of the components.

The procedures in this chapter cover disassembling and reinstalling door panels.

9.1.1.2 Related Information

For information about routine operational testing, see Chapter 5 of the manual *Testing and Troubleshooting* in this volume, part number TT1-500 093. For detailed troubleshooting techniques that address specific side door problems, see Chapter 18 of the same manual.

9.1.2 Safety Guidelines

9.1.2.1 Basic Precautions

All maintenance personnel must wear approved protective clothing and follow the safety guidelines outlined in the *Work Safely* booklet at all times.

9.1.2.2 Additional Safety Measures

Additional safety measures must be observed when working on the electrical and pneumatic systems of the side doors. Follow these rules in particular:

- Turn off all power to high-voltage equipment, and ground all wires before performing maintenance procedures.
- Always work with your assigned "buddy" or another technician present who can perform CPR or call for aid if necessary.

Alpha Draft Page 1

3. Choose **Structure>Structure View** to display the **Structure View** pane.

   The **Structure View** displays the document contents in a hierarchy of elements.

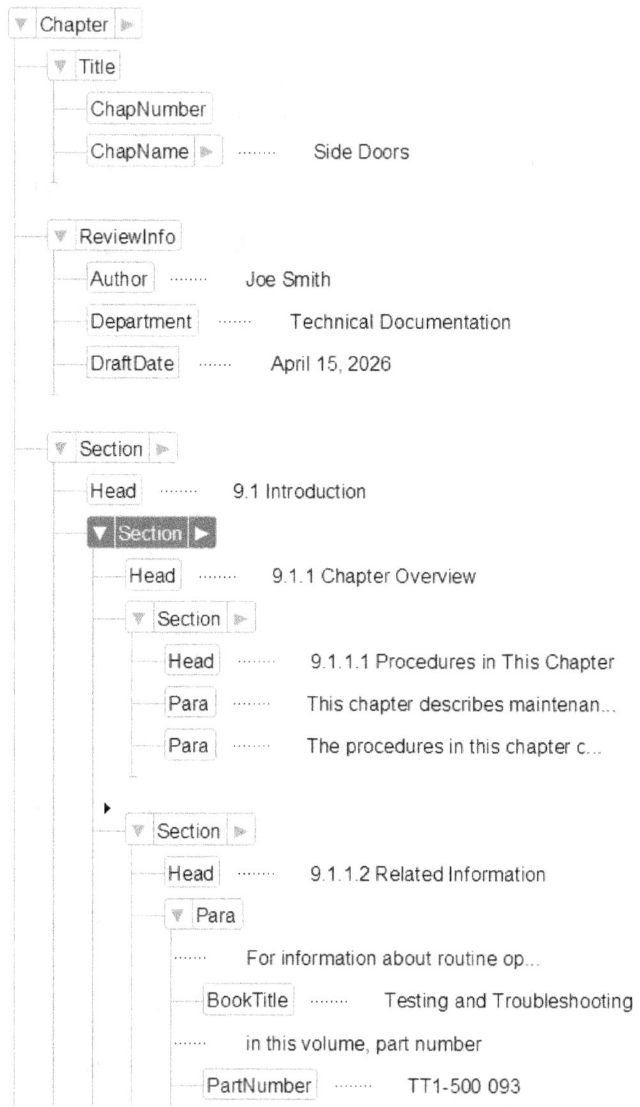

4. Choose **View>Panels>Element Catalog** to display the **Elements** panel.

   The **Elements** panel currently displays all the elements available for insertion in this document.

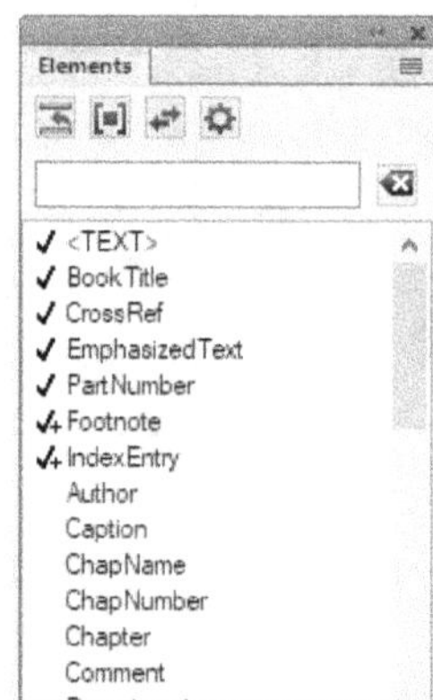

5. Create a new EDD based on the current document's element catalog by choosing **Structure>EDD>Export Element Catalog as EDD.**

EDD Version is 18.0

\\vmware-host\Shared Folders\Documents\Books\Fm v18-EDD WB\Workbook Files Master\EDDClassFiles\EDD-WB-Files-2026-18-0-0a\fm-sample.fm

April 15, 2026

Element (Container): Author
General rule: <TEXT>

Text format rules

1. **In all contexts.**
   Default font properties
   Language: None

Element (Container): BookTitle
General rule: <TEXTONLY>

Text format rules

1. **In all contexts.**
   Text range.
   Use character format: BookTitle

Element (Container): Caption
General rule: <TEXT>
Inclusions: IndexEntry, Footnote

Text format rules

1. **In all contexts.**
   Use format change list: Head

2. **In all contexts.**
   Basic properties
   Paragraph spacing
   Change space below by: 12.0 pt
   Default font properties
   Change size by: -1.0 pt
   Numbering properties
   Autonumber format: F:Figure 9.<n+>
   Character format: CaptionNumber
   Advanced properties
   Frame below: FigureCaption

An untitled FrameMaker document appears (an EDD), with the element and attribute definitions corresponding to **Elements** panel of the Sample.fm structured document. Note this document refers to the file from which it was exported.

6. Scroll through the EDD, viewing its contents.

   Take note of the **Structure View** and the elements displayed there.

   Note the addition of formatting specifications in this file.

   Next, you will create an EDD from scratch. Because, you are experimenting with the various ways to create an EDD, you will not save this EDD.

7. Close the untitled EDD without saving.
8. Close the **fm-sample.fm** document without saving.

Exercise 5: Creating a new EDD from scratch

In this exercise, you will create a new EDD and save it in your Class directory with an appropriate filename.

1. Choose **Structure>EDD> New EDD.**

   The new EDD appears.

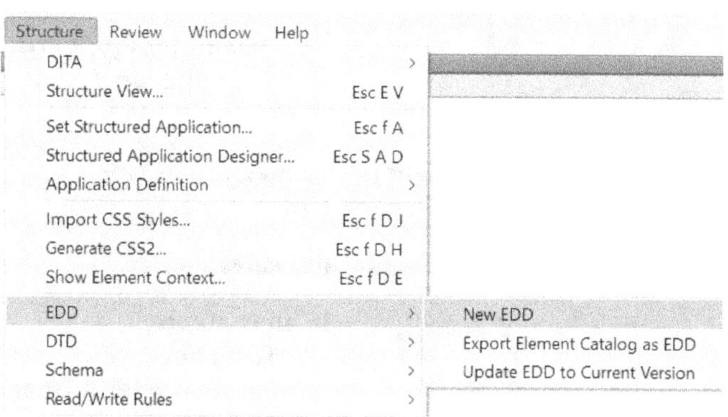

2. If not already open, open the **Structure View** and the **Elements** panel.

   The **Structure View** displays the elements automatically inserted with any new EDD, with the insertion point to the right of the Tag element bubble. If needed, click to the right of the tag element in the **Structure View.**

   The **Elements** panel displays **<TEXT>,** prompting you to type the Tag of the first element you are defining.

   Next, you will save the EDD with an appropriate filename.
   After that, you will analyze a document for its structure.

   This is a critical step before beginning to define your elements.

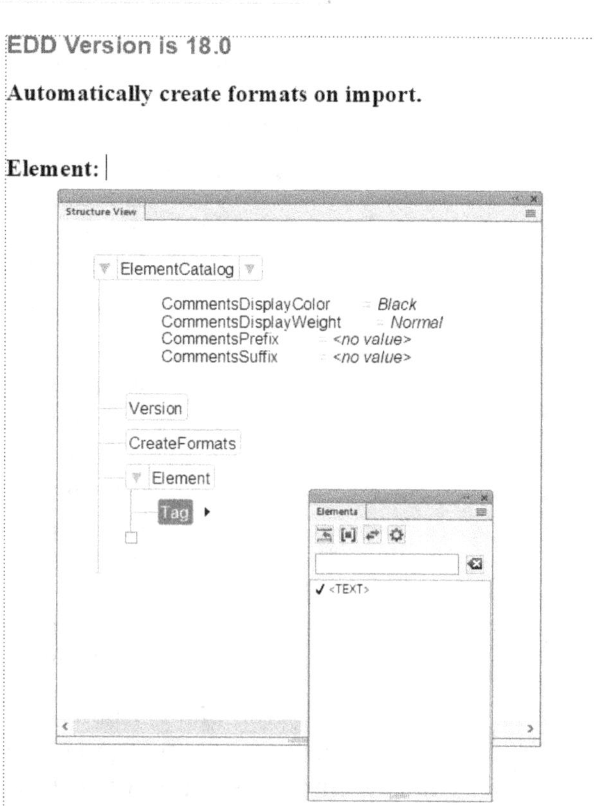

3. Use the **File>Save As** dialog to save the EDD in the directory containing your class files with the new filename, `EDD.fm`.
    a. From the File menu, choose **Save As.**

       The **Save Document** dialog appears.
    b. If necessary, change to your class files directory.
    c. In the **File name** field, delete the current file name and type: `EDD.fm`
    d. Click **Save.**

If you omit the .fm extension, FrameMaker will automatically add it for you.

Visualizing a content model

Throughout the EDD development in this training course, you will be acting as the structure developer for the publications department of a fictional transit authority.

You will ensure that all maintenance manuals written by your department conform to a consistent structure model.

Looking ahead

For the rest of the course, you will:

- Analyze a sample document type for its structure
- Define elements and attributes to be defined in an EDD
- Test the element definitions by
    - Importing them into a template
    - Inserting, wrapping, changing, merging, splitting, and unwrapping elements
    - Inserting and editing attribute values

Document analysis

When you analyze your document type (in this case, maintenance manuals), it is helpful to draw a diagram of the document's structure, showing:

- Logical chunks of information, called elements
- Element hierarchy
- Element content
- Element order
- Notations about frequency, and whether the elements are required or optional
- Attribute information

The structure diagram can be as minimal or as detailed as you want.

Typically, the more analysis done up front, the easier it will be to define your elements.

When analyzing your content:

- Work as a small team to get input from all those involved in using and approving the structure—authors (end users), format designers, structure developers, managers
- Work with a representative sample of the content
- Incorporate any company specifications—structure or style guides
- Analyze from the top down—bigger chunks first, then smaller chunks as needed
- Provide only as much structure as needed—don't over analyze

If after learning this material you have (or are transitioning to) an existing content model, then the content analysis will have already been completed.

If you will be creating a custom model for your internal use, then the content analysis will be a very important part of your migration project.

Exercise 6: Analyzing your content

In this exercise, you will analyze your content for its structure, the most important step before beginning to define your elements.

Normally, you would be viewing many samples of chapters, as well as the entire manual. In this set of materials you will start with something simple and enhance its structure as you go along.

1. Review the sample document provided (the next six pages of this module) of a chapter in the maintenance manual for your fictional organization.

   You will finish by defining an entire book with a table of contents and index by the end of the course.

2. For practice, take out a sheet of paper and take about five minutes to map out a diagram of its structure as you see it, similar to the diagram on page 17 in this module.

Normally, this is an iterative process, involving a team of authors (end users), format designers, structure developers, and managers. It might take hours, if not days or even weeks of revision for more complicated structures.

3. If you are learning this material in a class or course, compare your structure diagram to your neighbor's, noticing the different possibilities for this fairly simple structure. If not, consider how you might alternately represent the content.
4. Compare your structure diagram to the one on page 17, familiarizing yourself with its:
   - Element tags
   - Element hierarchy
   - Element content
   - Element order
   - Notations about frequency and whether the elements are required or optional

Once you're finished with this exercise, review the reference tables at the end of the module. They describe the relationship between different types of elements and their component parts. You may want to refer back to these tables as you go through the material.

Hardcopy Sample for Structure Diagram

Page 1

Chapter 1. Side Doors

1.1. Introduction

1.1.1. Chapter Overview

1.1.1.1. Procedures in This Chapter

This chapter describes maintenance procedures for the side doors on the AstroLiner T440B and T442 light rail cars. It includes safety guidelines, an overview of door components, and a maintenance schedule for some of the components.

The procedures in this chapter cover disassembling and reinstalling door panels.

1.1.1.2. Related Information

For information about routine operational testing, see *Chapter 5 of the manual Testing and Troubleshooting* in this volume, part number TT1-500 093. For detailed troubleshooting techniques that address specific side door problems, see *Chapter 18 of the same manual.*

Page 2

1.1.2. Safety Guidelines

1.1.2.1. Basic Precautions

All maintenance personnel must wear approved protective clothing and follow the safety guidelines outlined in the *Work Safely booklet* at all times.

1.1.2.2. Additional Safety Measures

Additional safety measures must be observed when working on the electrical and pneumatic systems of the side doors. Follow these rules in particular:

- Turn off all power to high-voltage equipment, and ground all wires before performing maintenance procedures.
- Always work with your assigned "buddy" or another technician present who can perform CPR or call for aid if necessary.

Some of the procedures in this chapter have warnings regarding the hazards associated with specific tasks. Failure to observe these warnings may result in injury or fatality.

1.1.3. Maintenance Overview

1.1.3.1. Components of the Side Doors

All AstroLiner T440B and T442 light rail cars have six sets of top-hung side doors. See Figure 1. Components of the side doors, on page 2. Each door set consists of a panel that slides to the right and a panel that slides to the left, operated by a pneumatic door operator.

Each set of doors has its own emergency release unit with a warning bell assembly. Each individual panel has an obstruction-detecting sensitive edge and a bottom brush seal.

Figure 1. Components of the side doors

1.1.3.2. Maintenance Schedules

Comprehensive overhaul procedures are scheduled semi-annually. Barring malfunction needing immediate attention, doors are scheduled for weekly operational checks and maintenance is performed as necessary. See Table 1. Items needing routine maintenance, on page 2. The table lists the items that should be checked each week.

Table 1. Items needing routine maintenance

Item	Part number	Per car
Door operator	83-48	4
Emergency release unit	36-95	2
Warning bell assem-bly	41-91	1
Signal bell assembly	78-90	1
Brush seals	46-94	8

1.1.3.3. Tools and Materials Required

Make sure that you have the following tools and materials before beginning any maintenance tasks.

- Screwdrivers

Flat-head and Phillips-head

- Adjustable wrenches
- Approved lubricant—HS1200 high-speed bearing grease
- Voltage meter
- Air pressure gauge
- Approved blow dryer—Model 390 or 490

1.2. Procedures

1.2.1. Door Panel Removal

1.2.1.1. When to Remove Door Panels

Remove door panels under the following conditions:

- when a door panel is physically damaged and needs to be replaced or repaired
- when a panel is jammed and approved techniques for loosening it are unsuccessful

If operations checks indicate that electrical power is not reaching the doors, or that the pneumatic system is not functioning properly, conduct additional tests on the door opera-

tor and repair that component as necessary.

1.2.1.2. Procedure

The following procedure applies to removal of one or both door panels on a side door. As you remove the panels place all small parts on clean rags, using one rag for each type of part.

1. Unlock and lift the access panel covering the hanger assembly.

   See Figure 2. Hanger assembly, on page 4. The figure shows a side view of the assembly, without the access panel.

Figure 2. Hanger assembly

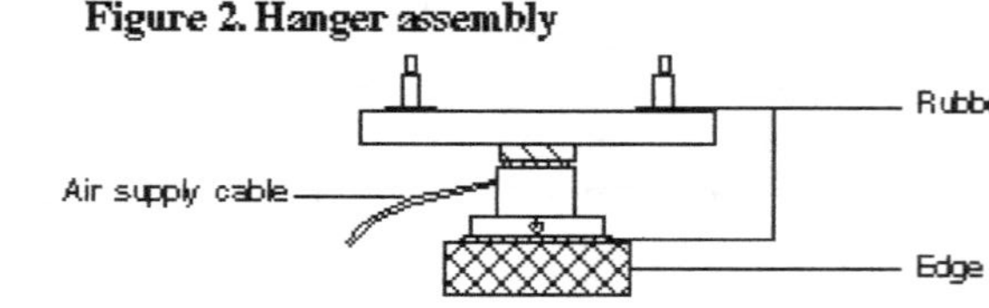

2. Unplug the air supply cable.
3. Loosen and remove the hex nuts and lock washers.
4. Remove the rubber seals.

Structure Diagram

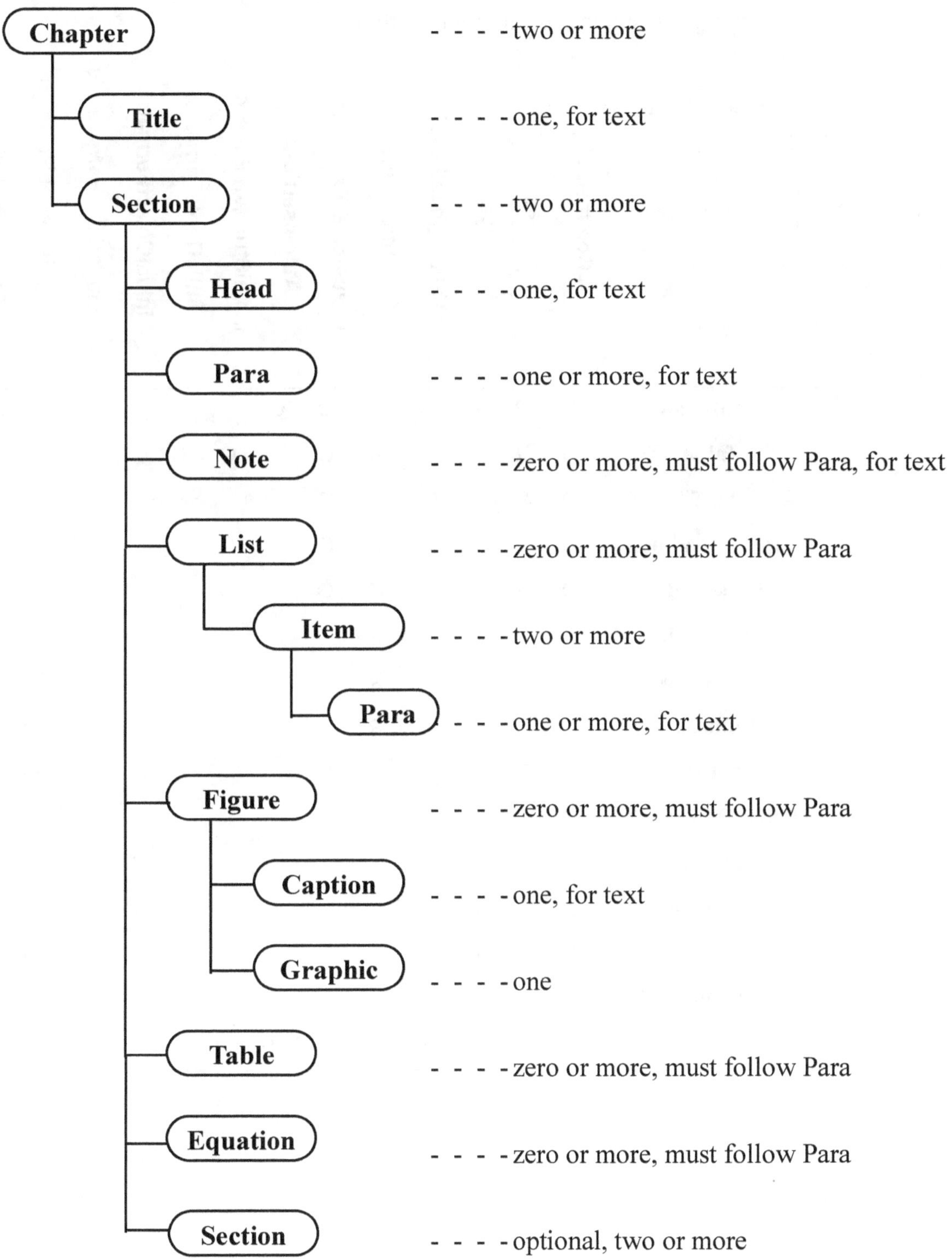

Relationship between EDD Elements and their component parts

An EDD itself has a content model applied to it. For each type of element defined in an EDD, there are required and optional elements available. The following chart shows which objects allow or require certain other objects within their definition.

	Comments	Tag	Type	ValidHighestLevel	GeneralRule	Inclusion & Exclusion	AutoInsertions	InitialStructurePattern	InitialTableFormat	InitialObjectFormat	SystemVariableFormatRule	AttributeList	TextFormatRules	First/LastParagraphRules	Prefix/SuffixRules
Container	O	R	R	R	R	O	O					O	O	O	O
Table	O	R	R		R	O		O	O			O	O		
TableTitle	O	R	R		R	O						O	O		
TableHeading	O	R	R		R	O		O				O	O		
TableBody	O	R	R		R	O		O				O	O		
TableFooting	O	R	R		R	O		O				O	O		
TableRow	O	R	R		R	O		O				O	O		
TableCell	O	R	R		R	O						O	O		
Footnote	O	R	R		R	O						O	O		
CrossReference	O	R	R							O		O			
Equation	O	R	R							O		O			
Graphic	O	R	R							O		O			
Marker	O	R	R							O		O			
SystemVariable	O	R	R								O	O			

R=Required, O=Optional

Basic Types of Elements You Can Define

Grouping	Type	Purpose
Containers, Tables and Table Parts, Footnotes	**Container**	General-purpose elements for text, child elements, or both
	Table	Parent of entire table
	TableTitle	Contains text for title of table
	TableHeading	Contains 1+ rows
	TableBody	Contains 1+ rows
	TableFooting	Contains 1+ rows
	TableRow	Contains 1+ cells
	TableCell	Contains text and/or child elements
	Footnote	Appears at bottom of column
Objects	**CrossReference**	For referencing other elements
	Equation	For inserting equations
	Graphic	For holding graphics
	Marker	For holding index or other markers
	SystemVariable	For inserting date, filename, or other system variables

Chapter 2: Initial Definition of Your First Element

Introduction

Now that you have created and saved your EDD, you'll need to expand on the rules it will contain.

The following list shows the various parts of an element definition. It is a list of the various things an element definition might contain:

- Comments
- Tag
- Type
- ValidHighestLevel
- GeneralRule
- Inclusion & Exclusion
- AutoInsertions
- InitialStructurePattern
- InitialTableFormat
- InitialObjectFormat
- SystemVariableFormatRule
- AttributeList
- TextFormatRules
- First/LastParagraphRules
- Prefix/SuffixRules
- Container

Each of these things will have its own rules about what elements they might contain as well.

This module focuses on the preliminary parts: the element's **Tag**, **Comments**, **Type**, and **ValidHighestLevel** settings for your Chapter element.

You'll learn about other options for this highest level element and for other elements in subsequent chapters.

Objectives

- Set **Elements** panel and other options to improve navigation
- Review purpose of preliminary element parts: **Tag**, **Comments**, **Type**, and **ValidHighestLevel**
- Insert the **Element** element
- Give the element its **Tag**
- Insert the **Comments** and type a descriptive element comment
- Select the **Type** of element
- Specify that an element is **ValidHighestLevel** in its flow

Setting up your editing environment

Exercise 1: Opening your EDD

You may still have `EDD.fm` open from the previous chapter. If not, open it from your class files directory.

If you didn't complete the Chapter 1 exercises, open **Chapter 2-start Initial EDD.fm** from your class files directory and resave it as `EDD.fm`.

For instructions on downloading class files, see "Downloading class files" on page 1.

The following two exercises are optional, but will allow you to more rapidly and intuitively complete the remainder of this workbook.

Exercise 2: Setting Toolbars

There are a number of toolbars that are shown by default in FrameMaker.

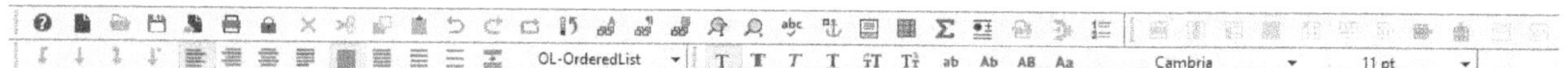

While you may find some useful, many are designed for standard unstructured FrameMaker.

In this exercise you'll hide all toolbars and then display only the Structured Access toolbar (not shown by default), which will give you quick access to many of the things you'll want to use in structured FrameMaker.

1. Navigate to **View>Toolbars>Hide All.**

   Note the reduction in clutter at the top of the screen.

2. Navigate to **View>Toolbars>Structured Access Bar.**

   The Structured Access bar appears, giving you an easy way to access the following:

   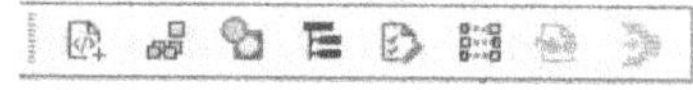

   - **XML View** (only active when editing XML, not SGML documents)
   - **Element Catalog** (brings up the **Elements** panel)
   - **Edit Attributes**
   - **Structure View**
   - **Validate**
   - **View Element Boundaries (as tags)**
   - Preview Filter by Attribute
   - Pretty Printing

You now have more room to increase the zoom of the content on your screen, and you also have less clutter to move through as you work.

Feel free to show or hide any toolbars you like, but the toolbars you hid will not be referenced in this book.

Setting your FrameMaker environment as shown in the next 3 exercises will help you move more quickly through the material, especially if you are comfortable navigating the structure view with your cursor keys and using the Smart Insert for Elements function (Ctrl+1)

Exercise 3: Set Elements panel options

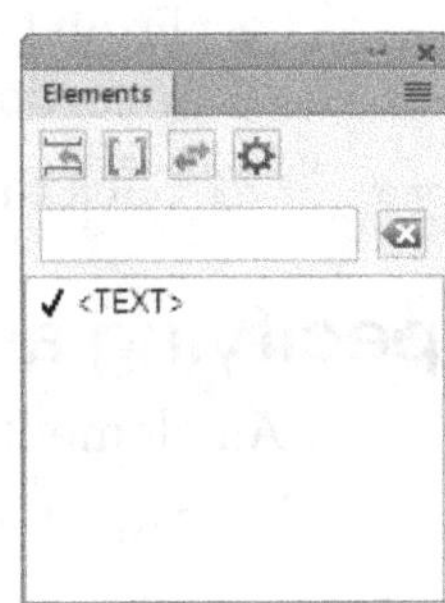

An EDD follows its own content model, so setting the **Elements** panel to display only the available elements will speed up your entry and will also help you understand the EDD content model itself.

1. (if necessary) Click the **Element Catalog** button in the **Structured Access** toolbar to bring up the **Elements** Panel.
2. In the **Elements** panel, select the **Settings** button (⚙) to display the **Set Available Elements** dialog.

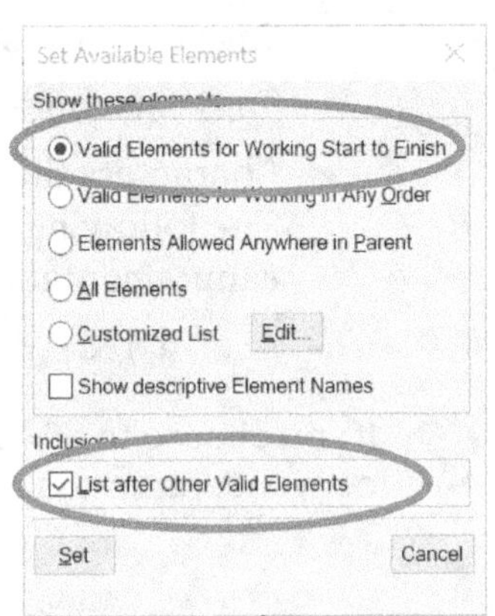

3. Select the following options:
   - Show these elements:
     Valid Elements for Working Start to Finish
   - Inclusions:
     List after Other Valid Elements
4. Select the **Set** button.

Your elements panel will now show a subset of elements, based on your current location within a document, and will show valid but infrequently used elements (when defined as **Inclusions**) after the more frequently used elements.

In some content models (like DITA) the **Show descriptive Element Names** option will initially help users understand the naming conventions used to label elements. The elements you'll define in this book use fairly obvious naming conventions, so enabling this option isn't critical to your success in working through the lessons.

Exercise 4: View Element Boundaries

You may find that navigating within the EDD is easier when viewing the element boundaries. Viewing boundaries allows you to move more accurately within the structure view with your left and right cursor arrows.

1. Select **View>Element Boundaries** (if element boundaries are not already visible)

You should now see square brackets as shown here.

If your text symbols are visible, use the **View** menu to turn off paragraph marks, or pilcrows, as they are not useful when working with structured documents and not useful in EDDs in particular.

[**Automatically create formats on import.**]

[[**Element:**]]]

Exercise 5: Set New Element Options

You will significantly reduce the effort needed to insert new structure by setting behavior of new elements as shown here.

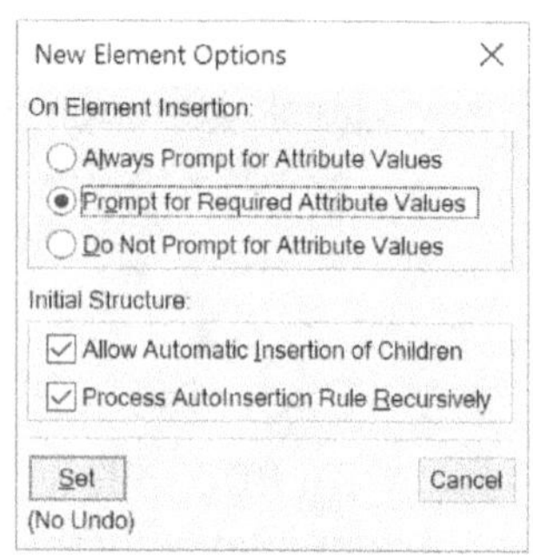

1. Navigate to **Element>New Element Options.**
2. From the **On Element Insertion**: area, select the **Prompt for Required Attribute Values** option.
3. From the **Initial Structure**: section, select the **Allow Automatic Insertion of Children** option.
4. Click the **Set** button.

Specifying an Element Tag

An element tag is a required part of all element definitions

- An element tag is the label of an element and is what displays in the **Elements** panel
- A tag may contain up to 255 characters, but it's better to keep tags concise
- Tags are case-sensitive
- Tags can contain white space
- Characters that perform special functions in SGML or XML are not valid in Tag names. The following characters will be used within the EDD for other functions, so an element tag cannot contain any of these special characters:

  ( ) & | , * + ? < > % [ ] = ! ; : { } "

If you choose to mix case and use white space in element names, strive for consistency in your conventions across all the named elements in your content model.

Exercise 6: Typing the Element Tag

In a new EDD, the first **Element** element and child **Tag** element are inserted automatically.

In this exercise, you will type the **Tag** of the first element, naming it `Chapter`.

1. Open the file `EDD.fm` that you created in the previous chapter.
2. In the **Structure View** of the `EDD.fm` file, click to the right of the **Tag** element.
3. In the Tag element, type: `Chapter`
4. Save your document.

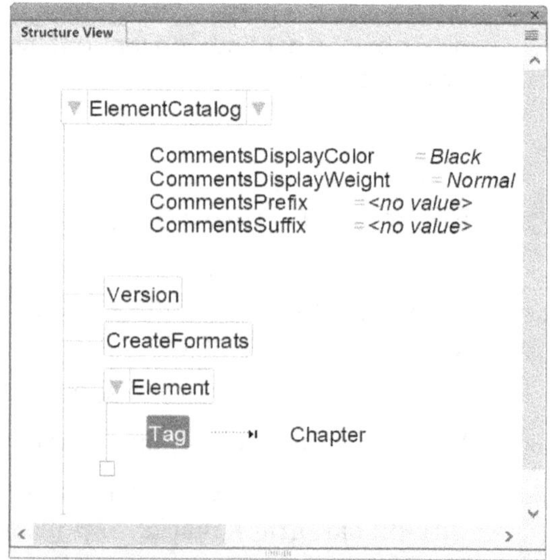

Specifying Element Type

Required part of all element definitions

Some elements match up to special parts, like cross-references or index markers, while other elements simply contain text or other elements. Elements containing text or other elements are generally defined as **Containers**.

Exercise 7: Identifying the Element as a Container

In this exercise, you will identify **Chapter** as a **Container** element. To do this, click below the **Tag** element, in the **Chapter** element definition.

1. In the **Structure View**, click as shown below the **Tag** element, but above the red square at the end of the **Element** element.

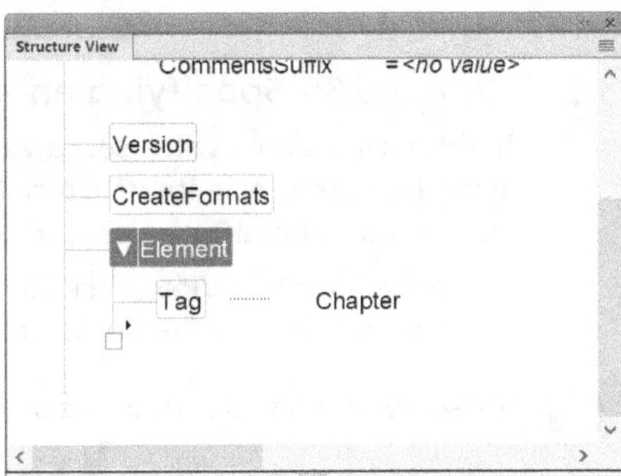

When positioning the cursor in the structure view, practice clicking where the triangle will appear. You'll be less likely to improperly position the cursor if you can predict (and click on) the area where the triangle will display.

2. From the **Elements** panel, double click on **Container** to insert a **Container** element.

   The **Container** element and **GeneralRule** child elements appear automatically.

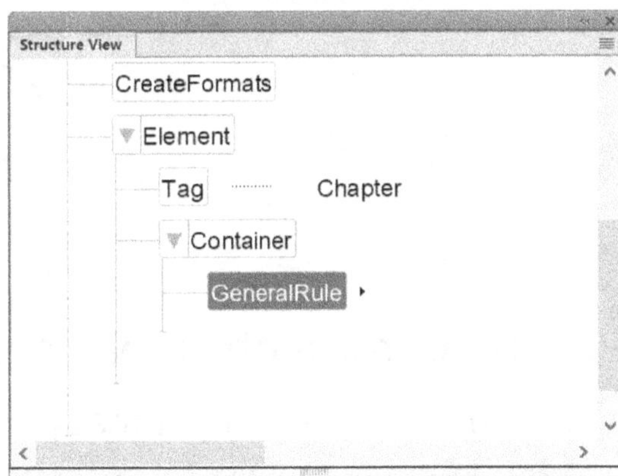

If the **General Rule** element doesn't appear automatically, set **Element>New Element Options** to **Allow Automatic Insertion of Children.**

   You'll learn about and define the **GeneralRule** in the next chapter.

3. Save your changes.

Specifying a starting element for a document

ValidHighestlevel is a required part of at least one container element definition for each structured flow

- It identifies a container that will hold all elements in flow
- It is required for the highest-level element for book files
- The **ValidHighestLevel** element can appear above or below a general rule

Exercise 8: Specifying an element as ValidHighestLevel

In this exercise, you will insert a **ValidHighestLevel** element to define the **Chapter** element as valid at the highest level in its structured flow. A **ValidHighestLevel** element may be placed either above or below a **GeneralRule** element.

1. In the **Structure View**, click above or below the **GeneralRule** of the **Chapter** element, on the line descending from the **Container** element.

The screen capture below shows the **ValidHighestLevel** placed **above** the **GeneralRule** element.

2. From the **Elements** panel, insert a **ValidHighestLevel** element.

   A **ValidHighestLevel** element and **Yes** child element appear.

3. Save your changes.

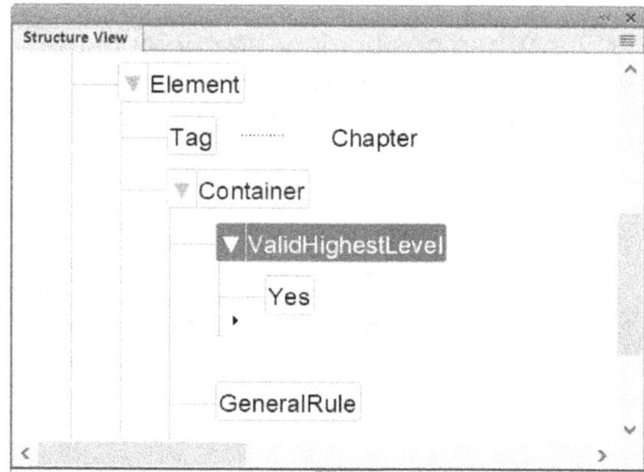

Placing comments in element definitions

Comments are an optional part of all element definitions

- Comments allow you to type a description of an element in your own words
- Comments may contain an unlimited number of characters
- They must be placed above the **Tag** element

Exercise 9: Commenting the Element Definition

In this exercise, you will insert a **Comments** element and add a comment describing your **Chapter** element.

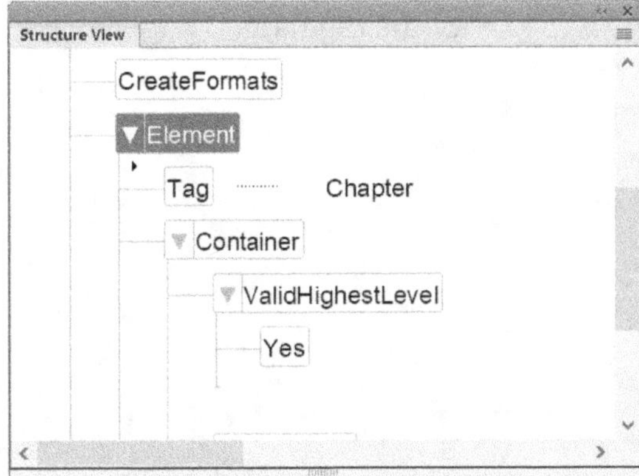

1. In the **Structure View**, click above the **Tag** element, on the line descending from **Element** element.

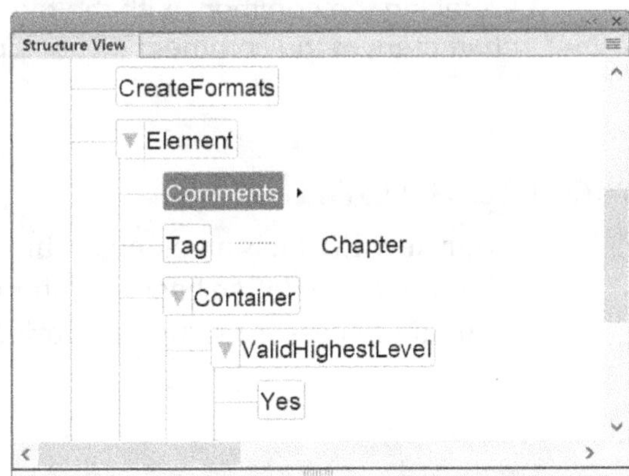

2. Double-click on the **Comments** element in the **Elements** panel.

   A **Comments** element appears.

3. In the **Comments** element, type:

```
Chapter is the container for all elements within chapters of
maintenance manuals
```

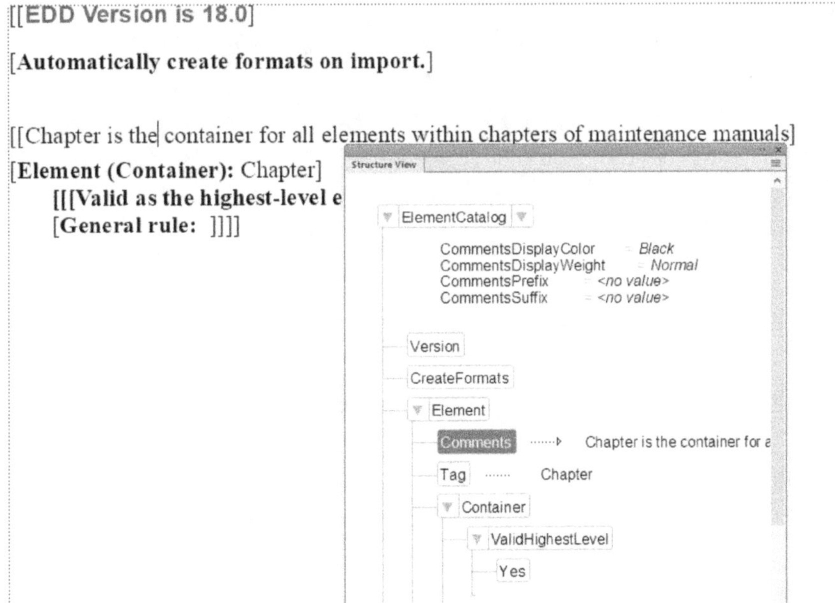

4. Collapse the element to see that the comment snippet shows now instead of the tag name.

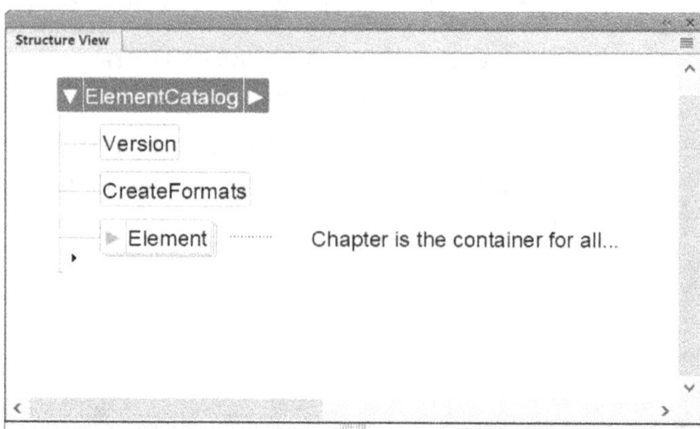

Prefacing the comment with the tag name is a good idea, since when collapsed, only the first few characters of the comment (the snippet) display, instead of the element's tag.

Moving forward

Your new EDD has an element, but now needs much more before you can start to use it. In the next chapter you'll define basic chapter components and see how your element definitions will evolve as you add necessary structure options.

Chapter 3: GeneralRule for Containers and Footnotes

Introduction

This chapter focuses on defining the **GeneralRule** for **Container** elements and **Footnote** elements.

Objectives

- Create new elements with a **GeneralRule** for **Container** elements and **Footnote** elements
- Learn about occurrence indicators and connectors in a **GeneralRule**
- Use content symbols and parentheses in a **GeneralRule**
- Review the default **GeneralRule** for **Container** and **Footnote** elements
- Import your EDD into a structured template
- Test your element definitions in the structured template

Overview

General rules are a required part of element definitions for:

- Containers
- Table, TableTitle, TableHeading, TableBody, TableFooting, TableRow, TableCell elements
- Footnotes

General rules specify:

- The child elements allowable in an element
- Whether child elements are required or optional
- The frequency in which child elements can occur
- The order in which child elements can occur
- Whether element may contain <TEXT>

Elements defined in your EDD must be referenced in at least one general rule.
Elements referenced in a general rule must be defined in your EDD.

Syntax

Refer to this page as needed to complete the exercises in this chapter.

Connectors

You can use connectors to separate multiple element tags in **GeneralRule** and to specify order of child elements

Here are the connectors allowable in a general rule:

Symbol	Meaning	Example
Comma (,)	Elements are mandatory and must occur in the specified order	TableTitle, TableHead, TableBody
Ampersand (&)	Elements are mandatory, but can occur in any order	Caption & Graphic
Vertical bar (|)	Any one element in the group can occur	Warning | Note | Caution

Occurence Indicators

Occurrence indicators allow you to specify whether a child is required or optional, and if it can be repeated

Here are the occurrence indicators allowable in a general rule:

Symbol	Meaning	Alternate Description
No indicator	Child is required and must occur only once	Requires 1
Question mark (?)	Child is optional and can occur once	May occur 0 or 1 time
Asterisk (*)	Child is optional and can occur more than once	May occur 0 or more times
Plus sign (+)	Child is required and can occur more than once	Must occur 1 or more times

Special Content Types

You can use content strings to specify content other than child elements and to indicate elements with no content

String	Meaning
<TEXT>	Can contain text and any inclusions
<TEXTONLY>	Can contain only text Cannot contain child elements, even inclusions defined in ancestors' content rules
<ANY>	Can contain any combination of text and elements defined in EDD
<EMPTY>	Cannot contain any text or elements

Writing the initial GeneralRules for basic elements

Exercise 1: Writing a GeneralRule for a Chapter element

In this exercise, you will define a **Chapter** as an element that will contain a **Title** element, and is followed by two or more **Section** elements.

1. If it is not already open, open `EDD.fm`, from the directory containing your class files.

   If you didn't complete the previous chapter exercises, open up the **Chapter 3-start Containers.fm** file in your class directory and save it as `EDD.fm` in your class files directory.

For instructions on downloading class files, see "Downloading class files" on page 1.

2. In the **Structure View**, click to the right of the **GeneralRule** element.
3. In the GeneralRule element, type:
   `Title, Section, Section+`

   Commas specify an explicit order. Spaces aren't essential but are helpful for readability.

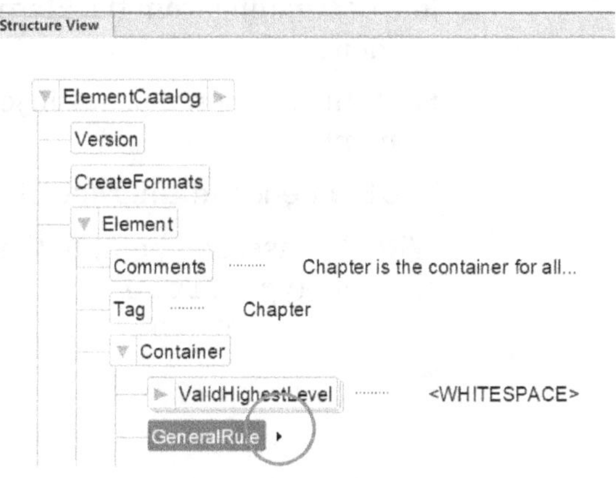

You might interpret this rule as "A valid chapter must contain a title, followed by a section, followed by one or more optional sections" or "A valid chapter must contain a title, followed by two or more sections."

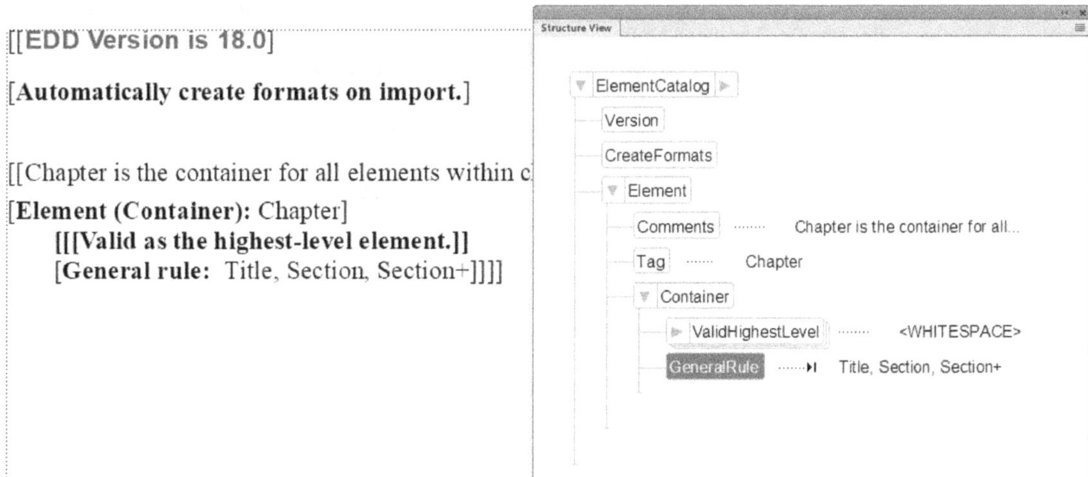

4. Save your changes.

Elements referenced in a General Rule must be defined. Elements defined in the EDD must be referenced in a General Rule, unless they are Valid at Highest Level or are Inclusions

Now that you have referenced a **Title** and a **Section** in a **GeneralRule**, you will need to define these elements.

Exercise 2: Defining a Title element

In this exercise you will create a new container element called **Title** with a **GeneralRule** of **<TEXT>**.

To add your title element, do the following:

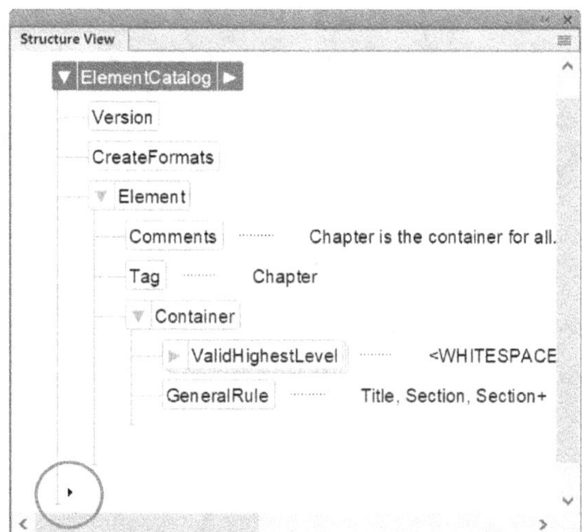

1. In the **Structure View**, click at the end of your EDD, below **Element** for the **Chapter** tag, on the line descending from the **ElementCatalog** element.
2. From the **Elements** panel, insert an **Element** element.

   The **Element** and child **Tag** element appear.
3. With the insertion point to the right of the **Tag** element, type: `Title`

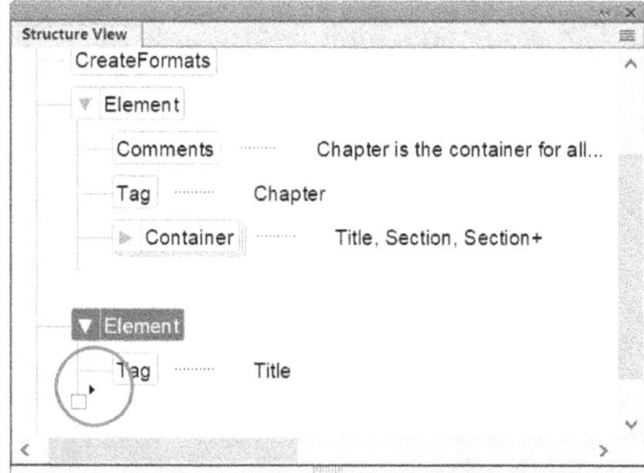

4. Click below the **Tag** element, above the red square beneath the **Element** element.

5. From the **Elements** panel, insert **Container**.

   The **Container** element and **GeneralRule** child element appear.
6. With the insertion point to the right of the **GeneralRule** element, type: **<TEXT>**
7. Save your changes.

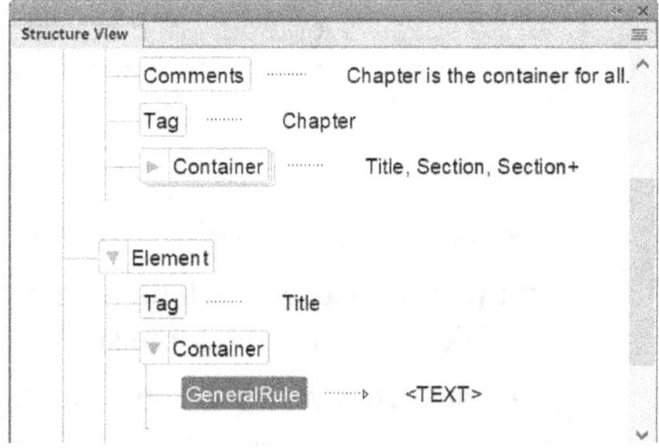

At this point you may want to start collapsing elements in your **Structure View** to improve navigation.

Exercise 3: Defining Section, a container of child elements

In this exercise, you will define the **Section** element that will (for now) require a **Head** element followed by one or more **Para** elements. The **GeneralRule** for the **Section** element will be one of the most frequently modified elements in this course.

To define your section element, do the following:

1. In the **Structure View**, click below the **Title** element definition, on the line descending from the **ElementCatalog** element.
2. Insert an **Element** from the **Elements** panel.

   The **Element** and its child **(Tag)** appear.
3. With the insertion point to the right of the **Tag** element, type: `Section`
4. Click below the **Tag** element, above the red square at the end of the **Element** element.
5. From the **Elements** panel, insert a **Container** element.

   Container and its child, **GeneralRule** appear. (not shown in screen capture)
6. With the insertion point to the right of the **GeneralRule** element, type: `Head, Para+`

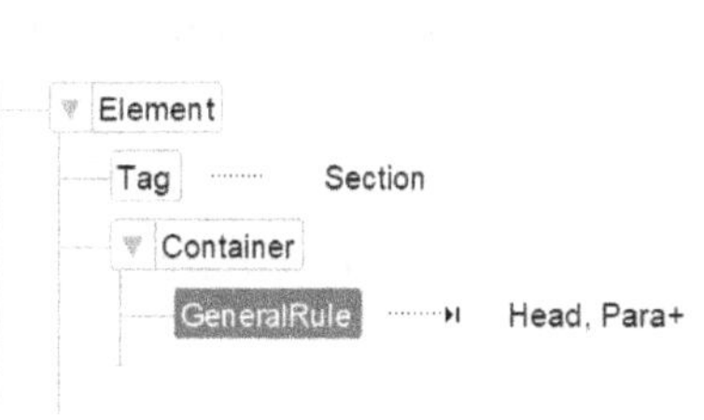

   The + indicates that Para is required, and can occur more than once.
7. Save your changes.

Next, you'll define the two newly specified elements, **Head** and **Para**.

White space is ignored when working with connectors. This means you are free to add white space in and around connectors to make the EDD more easily readable.

Exercise 4: Defining More Descendant Container Elements

In this exercise, you will define the **Head** and **Para** elements referenced in the previous exercise. If you like, you can match the screen captures shown for each element.

To insert your new elements, do the following:

1. Create a new **Element** as a sibling to **Section** and tag it **Head.**
    a. In the **Structure View**, click below the definition for the **Section** tag, on the line descending from the **ElementCatalog** element.
    b. From the **Elements** panel, insert an **Element** element.
       The **Element** and child **Tag** element appear.
    c. In the **Tag** element, type: `Head`
2. Define **Head** as a **Container** with a **GeneralRule** of `<TEXT>`
    a. Click below the **Tag** element.
    b. From the **Elements** panel, insert **Container.**
       The **Container** element and **GeneralRule** child element appear.
    c. In the **GeneralRule** element, type: `<TEXT>`
3. Create a new **Element** and tag it `Para`.
    a. In the **Structure View,** click below the last **Element** element.
    b. From the **Elements** panel, insert an **Element** element.
       The **Element** and child **Tag** element appear.
    c. In the **Tag** element, type: `Para`
4. Define **Para** as a **Container** with a **GeneralRule** of <TEXT>
    a. Click below the **Tag** element.
    b. From the **Elements** panel, insert **Container.**
       Container element and **GeneralRule** child element appear.
    c. In the **GeneralRule** element, type: `<TEXT>`
5. Save your changes.

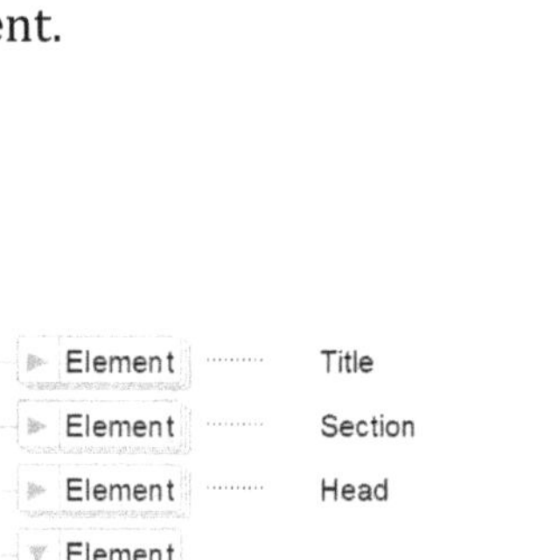

Importing and Testing

Exercise 5: Importing the EDD

In this exercise, you will create a new portrait document, save it as your structured template, then import the element definitions defined in the EDD into your structured template.

1. Create a new portrait document.
   a. Select **File>New>Document.**

      The **New** dialog box appears.
   b. Click **Portrait.**

      A new blank document appears.

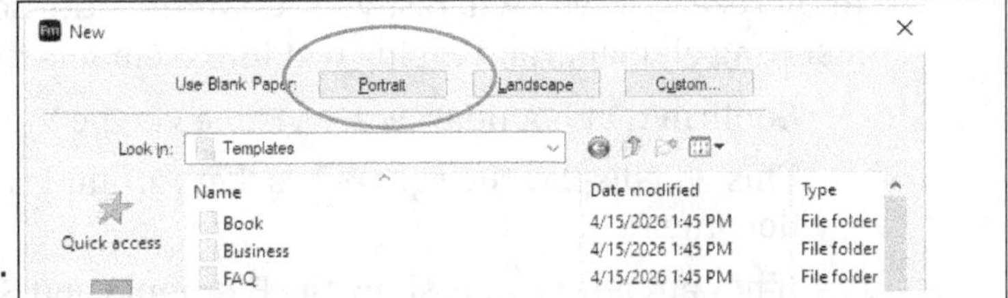

2. Use the **Save As** dialog box to save the new portrait document in your class files directory with the new filename `testdoc.fm`.
   a. From the **File** menu, choose **Save As.**

      The **Save Document** dialog box appears.
   b. If necessary, change to your class files directory.
   c. In the **Save in File** field, delete the current file name and type: `testdoc.fm`
   d. Click **Save.**
3. Ensure that your cursor is in the text flow and then look at the **Elements** panel.

   This is currently an unstructured document and the **Elements** panel is empty because there are no elements available for you to insert.
4. Ensure that `testdoc.fm` is the active file and choose **File>Import > Element Definitions.**

   The **Import Element Definitions** dialog box appears.
5. From the **Import from Document** dropdown menu, choose `EDD.fm`.

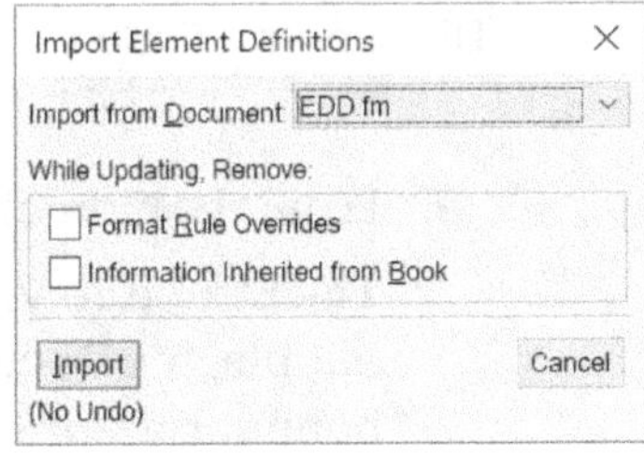

If the EDD isn't open in FrameMaker, it will not appear in the menu as a choice for import.

6. Click **Import.**

   An alert appears indicating completion of the import process.
7. Click **OK** to close the alert box.
8. If errors exist in the element definitions, a log file will appear identifying the problems.

   If this occurs, edit your EDD to correct the errors and repeat steps 3 through 6.

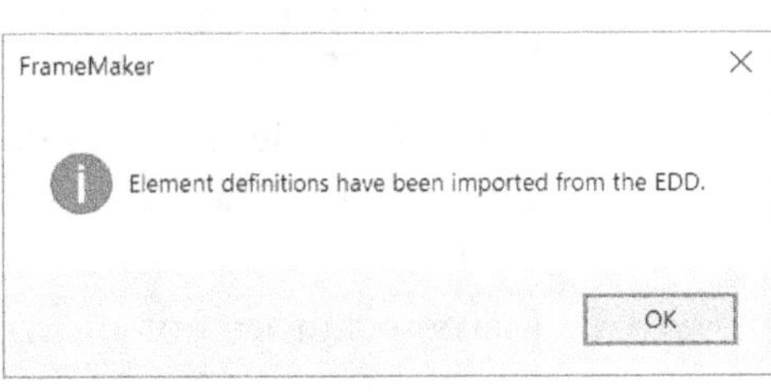

If no errors appear, FrameMaker has successfully added and/or updated the element definitions in the element catalog of `testdoc.fm`.

9. Save your changes.

Exercise 6: Testing the EDD

In this exercise, you will test your **Chapter**, **Title**, **Section**, **Head**, and **Para** element definitions by inserting elements and typing text into elements defined to contain text.

1. Confirm that boundaries are visible via the **View>Element Boundaries** command.

   This will display the square brackets around content once you've added elements to your document.

2. If not already visible, show the **Elements** and **Structure View** panels.

3. Set available elements to **Valid Elements for Working Start to Finish.**

   a. Choose **Element>Set Available Elements.**

      The **Set Available Elements** dialog box appears.

   b. In **Show these elements**, turn on **Valid Elements for Working Start to Finish.**

   c. In **Inclusions**, choose **List after Other Valid Elements.**

   d. Click **Set.**

      The **Elements** panel displays only the **Chapter** element as the element valid at the highest level in the structure.

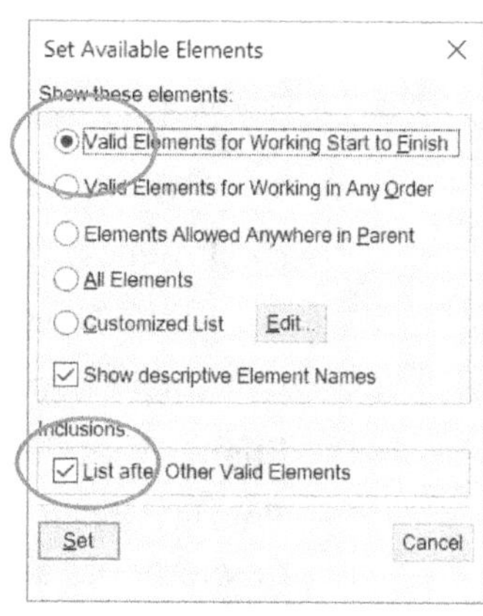

If you don't see **Chapter** in the **Elements** panel, place your cursor inside of the text frame in the Document Window.

4. From the **Elements** panel, insert a **Chapter** element.

   Chapter

   The **Structure View** displays the **Chapter** element bubble, the **Document Window** displays element boundaries for the **Chapter** element, and the **Elements** panel displays an available **Title** child element.

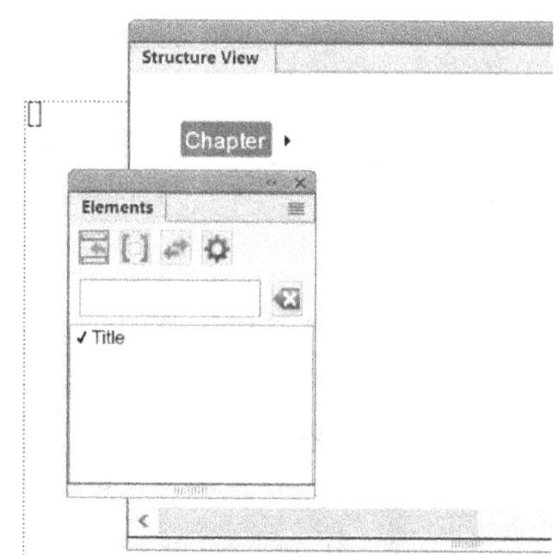

5. Insert a **Title** element.

   The **Elements** panel displays **<TEXT>**, indicating that you can type in this element.

6. In the **Title** element, type:
   `Title of Chapter in Maintenance Manual`

   (screen capture not shown)

7. Click below the **Title** element and insert a **Section** element.

   The **Elements** panel updates to reflect the new position of your cursor within the document.

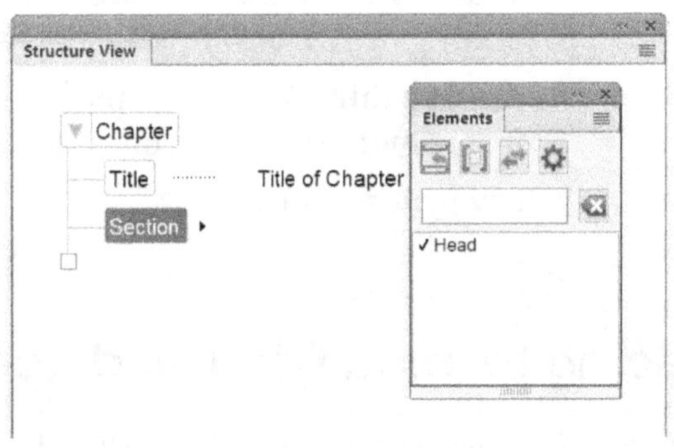

8. Insert a **Head** element and type: 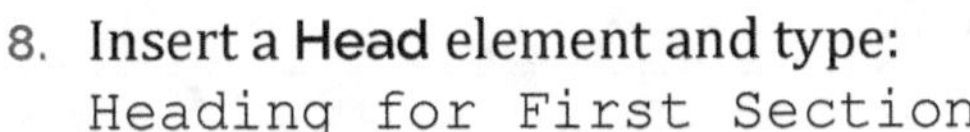
   `Heading for First Section`

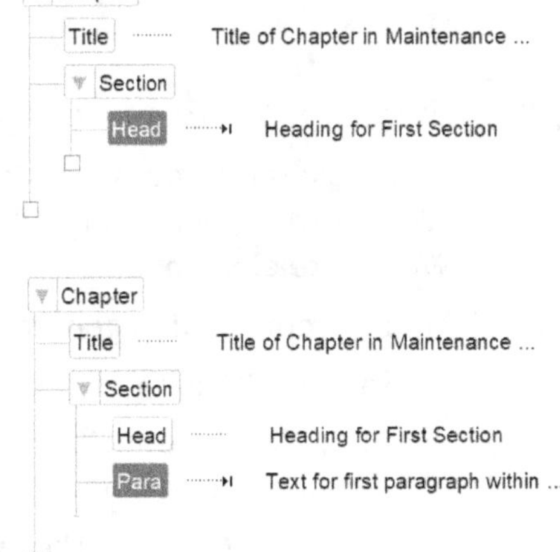

9. Click below the **Head** element, insert a **Para** element, and type: `Text for first paragraph within Section.`

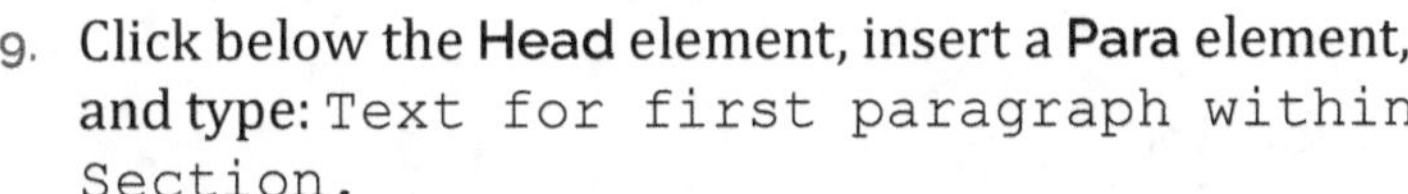

10. Click below the **Para** element and insert several more **Para** elements to verify that you can insert more than one **Para** element. (screen capture not shown)

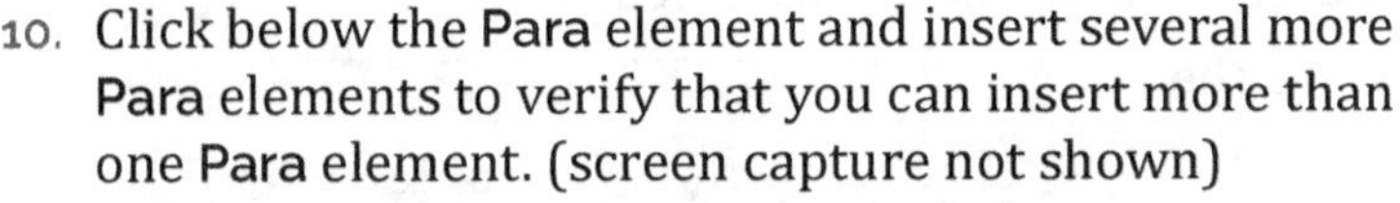

    Notice the red square on the line descending from the **Chapter** element.

    That's because **Chapter** is required to have two or more **Section** elements.

11. Insert another **Section** element, a **Head**, and at least two **Para** elements.

    Don't bother to type any more text. You are just testing, and you have already made sure that you can type in these elements when you inserted them in the first Section.

    Notice how the red square disappears once the **Section** requirement is satisfied.

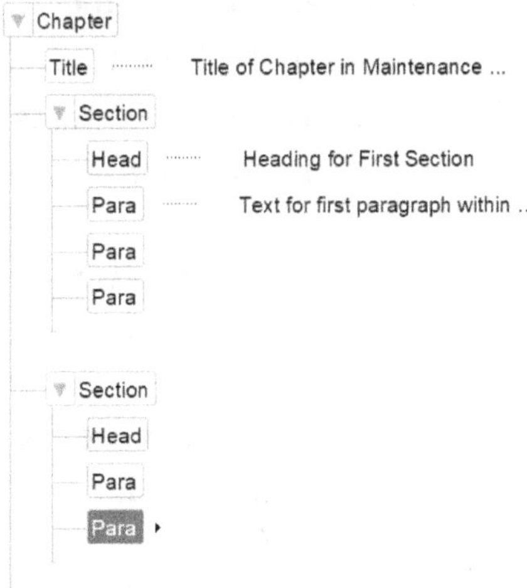

12. You may recall that the **GeneralRule** for **Chapter** allows for two *or more* **Section** elements.

    Insert a third **Section** element to verify that you can insert more than two consecutive **Section** elements.
13. Save your changes.

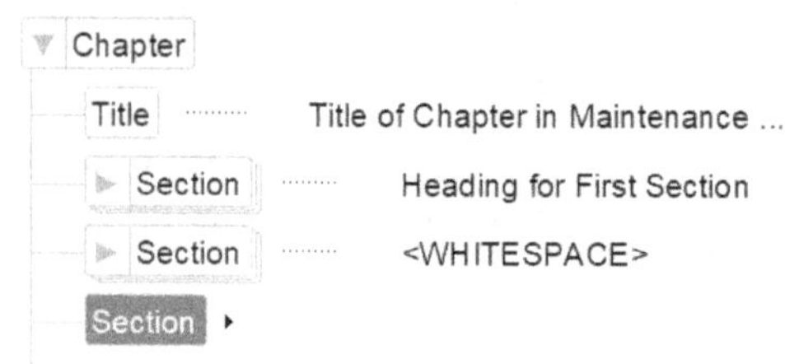

Adding to, modifying, and updating the EDD

From this point on, exercises will include changes to **GeneralRule** content in the EDD. The additional text for these definitions are marked in bold.

Exercise 7: Expanding the Section GeneralRule

In this exercise, you will redefine the **Section** element's **GeneralRule** to allow an optional **WarnNote** element that will contain `<TEXT>`. Then you will define the newly referenced **WarnNote** element.

Modify your **Section** general rule as follows:

1. In the EDD, locate the element definition for the **Section** element.
2. In the **Document Window**, change the **GeneralRule** as follows:

   `Head, (Para, `**`WarnNote?)`**`+`

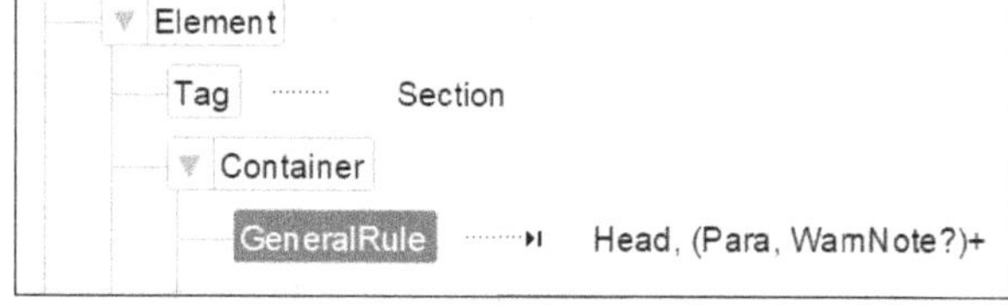

The newly referenced **WarnNote** element requires a definition.

3. Create a new **Element** and tag it **WarnNote.**
   a. In the **Structure View**, click below the last **Element** element.
   b. From the **Elements** panel, insert an **Element** element.

      An **Element** and a child **Tag** element appear.
   c. In the **Tag** element, type: `WarnNote`
4. Define **WarnNote** as a **Container** with a **GeneralRule** of <TEXT>
   a. Click below the **Tag** element.
   b. From the **Elements** panel, insert **Container.**

      Container element and **GeneralRule** child element appear.
   c. In the **GeneralRule** element, type: `<TEXT>`

You can use Copy/Paste to speed creation of similar structure.

5. Save your changes.

Exercise 8: Referencing a Group Within a Group

In this exercise, you will redefine **Section** to allow for more complex combinations of elements.

To modify the Section general rule, do the following:

1. In the EDD, locate the element definition for the **Section** element.
2. In the **Document Window**, change the **GeneralRule** as follows:

   `Head, (Para, (WarnNote | List)?)+`

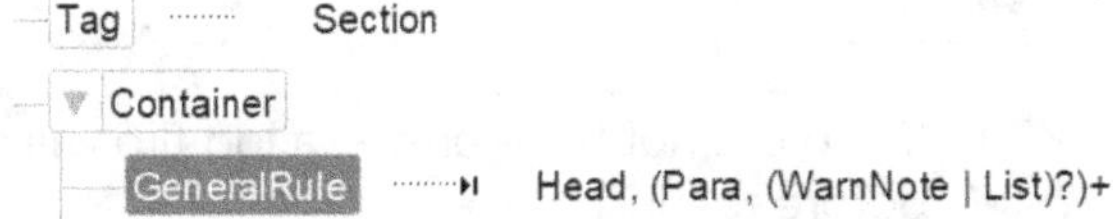

3. Create a new **Element** and tag it **List**.
   a. In the **Structure View**, click below the last **Element** element.
   b. From the **Elements** panel, insert an **Element** element.
      The **Element** and child **Tag** element appear.
   c. In the **Tag** element, type: `List`
4. Define **List** as a **Container** with a **GeneralRule** of **Item, Item+**
   a. Click below the **Tag** element.
   b. From the **Elements** panel, insert **Container.**
      A **Container** element and **GeneralRule** child element appear.
   c. In the **GeneralRule** element, type: `Item, Item+`
5. Create a new **Element** and tag it **Item**.
   a. In the **Structure View**, click below the last **Element** element.
   b. From the **Elements** panel, insert an **Element** element.
      The **Element** and child **Tag** element appear.
   c. In the Tag element, type: `Item`
6. Define **Item** as a **Container** with a **GeneralRule** of <TEXT>
   a. Click below the **Tag** element.
   b. From the **Elements** panel, insert **Container.**
      Container and child **GeneralRule** elements appear.
   c. In the **GeneralRule** element, type: `<TEXT>`

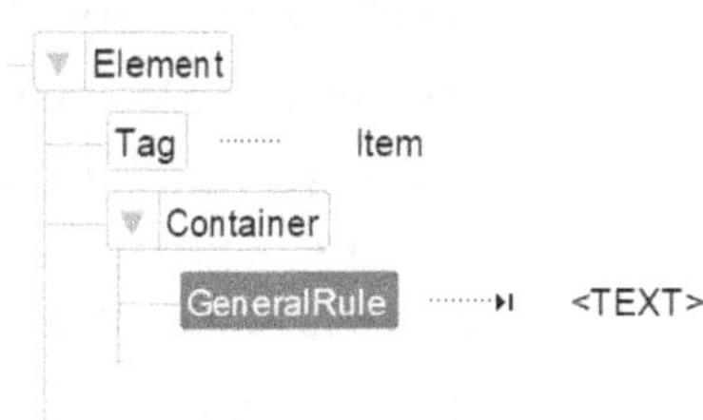

7. Save your changes.

Exercise 9: Reimporting and Retesting

In this exercise, you will reimport the EDD into the structured template and test your element definitions for **Section**, **WarnNote**, **List**, and **Item**.

1. Reimport your element definitions and fix any errors.
   a. In `testdoc.fm`, from the **File** menu, choose **Import > Element Definitions.**

The **Import Element Definitions** dialog appears.

b. From the **Import from Document** dropdown menu, choose `EDD.fm`.

c. Click **Import**.

An alert box appears indicating "**Element definitions have been imported from the EDD**"

d. Click **OK** to close the alert box.

If you had errors in the element definitions, edit your EDD and reimport.

If you did not have errors, a log file will not appear and you do not need to edit your EDD and reimport before testing.

FrameMaker replaces element definitions in the **Elements** panel of the structured template.

2. In the **Structure View**, click on the line descending from any **Section** element but below a **Para** element.
3. From the **Elements** panel, insert a **WarnNote** element and add text to test it.
4. Click on the line descending from any **Section** element but below a **Para** element.
5. Insert a **List** element.
6. In the **List** element, insert an **Item** element and type any text to test it.
7. Notice the red square on the line descending from the **List** element.

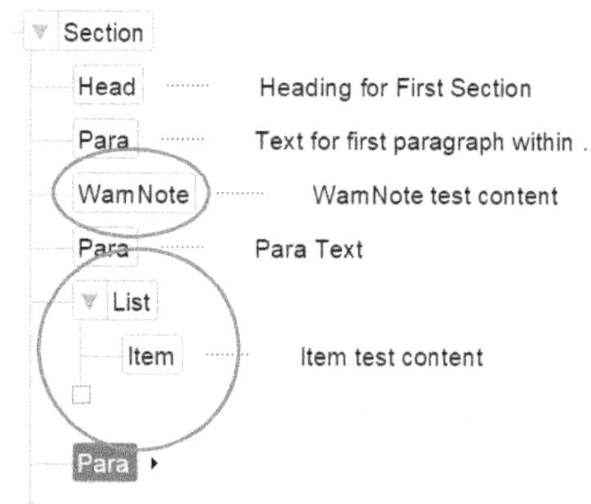

A **List** must have two or more **Item** elements to conform to the content model.

8. Insert several more **Item** elements to verify that you can have as many as needed.
9. Try inserting two consecutive **WarnNote** elements.

   Upon insertion of the second element a red square appears between the elements. This indicates that other content (in this case, a Para) is needed between two consecutive **WarnNote** elements.
10. Try inserting two consecutive **List** elements.

    Upon insertion of the second element a red square appears between the elements. This indicates that other content (in this case, a Para) is needed between two consecutive **List** elements.
11. Try inserting a **List** immediately after a **WarnNote**, or vice versa.

    Upon insertion of the second element a red square appears between the elements. This indicates that other content (in this case, a Para) is needed between **List** and **WarnNote** elements.

12. Try inserting a **List** after a **Head**.

    Upon insertion of the second element a red square appears between the elements. This indicates that other content (in this case, a Para) is needed between **Head** and **List** elements.

13. Save your changes.

Exercise 10: Referencing an Element Within Itself

In this exercise, you will redefine the **Section** element's **GeneralRule** to allow nested sections.

To do this:

1. In the EDD, locate the element definition for the **Section** element.
2. In the Document Window, change the **GeneralRule** as follows:

```
Head, (Para, (WarnNote | List)?)+, (Section, Section+)?
```

3. Save your changes.
4. Reimport your element definitions and fix any errors.
   a. In `testdoc.fm`, from the **File** menu, choose **Import > Element Definitions.**

      The **Import Element Definitions** dialog appears.

   b. From the **Import from Document** dropdown menu, choose `EDD.fm`.
   c. Click **Import.**

      An alert box appears indicating "**Element definitions have been imported from the EDD**"

   d. Click **OK** to close the alert box.

If you had errors in the element definitions, edit your EDD and reimport.

5. Practice inserting **Section** elements.

   Recognize that whenever you insert a **Section** element, you must insert two or more at the same level, and they must be the last two children of their parent.

6. Save your changes.

Exercise 11: Referencing and Defining a Footnote Element

In this exercise, you will redefine the **Para** element's **GeneralRule** to allow both text and footnote elements. Then, you will define a **Footnote** element that takes advantage of FrameMaker's existing footnote functionality.

To allow for footnotes, do the following:

1. In the EDD, locate the element definition for the **Para** element.
2. In the **Document Window**, change the **GeneralRule** as follows:

```
(<TEXT> | Footnote)+
```

3. Insert a new **Element** and tag it **Footnote.**
   a. In the **Structure View,** click below the last **Element** element.
   b. From the **Elements** panel, insert an **Element** element.
      An **Element** and a child **Tag** element appear.
   c. In the **Tag** element, type: `Footnote`
4. Define **Footnote** as a **Footnote** with a **GeneralRule** of <TEXT>
   a. Click below the **Tag** element.
   b. From the **Elements** panel, insert **Footnote.**
      Footnote element and **GeneralRule** child element appear.
   c. In the **GeneralRule** element, type: `<TEXT>`

Element
Tag Footnote
Footnote
GeneralRule <TEXT>

5. Save your changes.
6. Reimport the element definitions from your EDD into your test document. If you get errors when importing the element definitions, correct your EDD and reimport.
   Continue to the next step when your import occurs without errors.
7. In the **Structure View,** click on the line descending from any **Section** element.
8. From the **Elements** panel, insert a **Para** element and type any text to test it.
9. With your insertion point still in the text, insert a **Footnote** element and type any text to test it.
10. Click below the Footnote element, on the line descending from the Para element, and type a little more text.
11. With your insertion point still in the text, insert another **Footnote** element and type any text to verify that you can insert multiple **Footnote** elements in a **Para** element.
12. Try inserting a **Footnote** element in a **Head** element.
    You cannot. In your EDD, **General Rules** are currently written so that **Footnote** elements can be children of **Para** elements, not **Head** or other elements.
13. Save your changes.

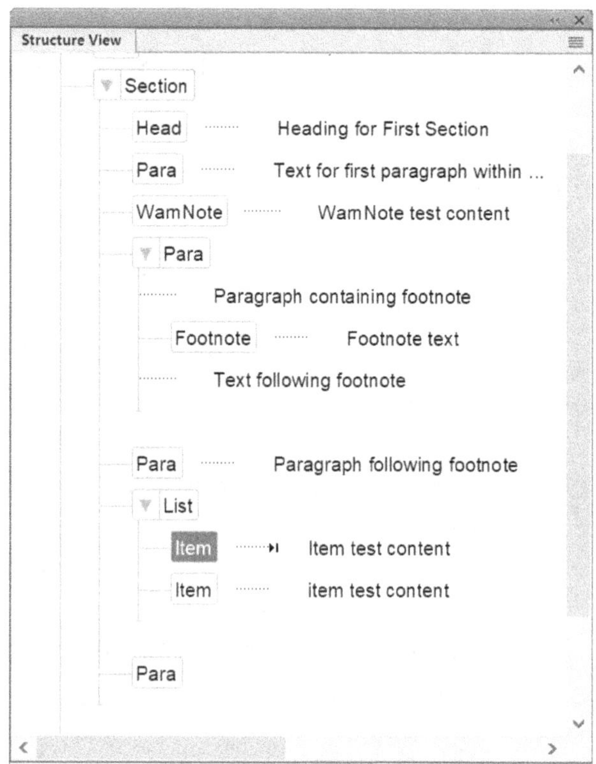

Chapter 4: GeneralRule for Tables and Table Parts

Introduction

This chapter focuses on defining the **GeneralRule** for **Table** elements and table part elements.

Objectives

- Write **GeneralRule** for **Table** element
- Write **GeneralRule** for table part elements—**TableTitle, TableHeading, TableBody, TableFooting, TableRow, TableCell**
- Review default **GeneralRules**
- Review restrictions on **GeneralRules** for **Table** and table part elements
- Reimport the EDD into the structured template
- Test **Table** and table part element definitions in the structured template

Overview

General rules are a required part of element definitions for:

- Containers
- Table, TableTitle, TableHeading, TableBody, TableFooting, TableRow, TableCell elements
- Footnotes

General rules specify:

- The child elements allowable in an element
- Whether child elements are required or optional
- The frequency in which child elements can occur
- The order in which child elements can occur
- Whether element may contain <TEXT>

Elements defined in your EDD must be referenced in at least one general rule.
Elements referenced in a general rule must be defined in your EDD.

Syntax

Refer to this page as needed to complete the exercises in this chapter.

Connectors

You can use connectors to separate multiple element tags in **GeneralRule** and to specify order of child elements

Here are the connectors allowable in a general rule:

Symbol	Meaning	Example
Comma (,)	Elements are mandatory and must occur in the specified order	TableTitle, TableHead, TableBody
Ampersand (&)	Elements are mandatory, but can occur in any order	Caption & Graphic
Vertical bar (|)	Any one element in the group can occur	Warning | Note | Caution

Occurence Indicators

Occurrence indicators allow you to specify whether a child is required or optional, and if it can be repeated

Here are the occurrence indicators allowable in a general rule:

Symbol	Meaning	Alternate Description
No indicator	Child is required and must occur only once	Requires 1
Question mark (?)	Child is optional and can occur once	May occur 0 or 1 time
Asterisk (*)	Child is optional and can occur more than once	May occur 0 or more times
Plus sign (+)	Child is required and can occur more than once	Must occur 1 or more times

Special Content Types

You can use content strings to specify content other than child elements and to indicate elements with no content

String	Meaning
<TEXT>	Can contain text and any inclusions
<TEXTONLY>	Can contain only text Cannot contain child elements, even inclusions defined in ancestors' content rules
<ANY>	Can contain any combination of text and elements defined in EDD
<EMPTY>	Cannot contain any text or elements

GeneralRule Restrictions

Element type	Restrictions
Table	• Limited to one each of TableTitle, TableHeading, TableBody, TableFooting (in that order) • TableBody is required • TableTitle, TableHeading, TableFooting are optional • No plus sign (+), asterisk (*), ampersand (&) • No <TEXT>, <TEXTONLY>, <ANY>, <EMPTY>
TableTitle	• All child elements are allowed, except Table and table parts • <TEXT>, <TEXTONLY>, <ANY>, <EMPTY> allowed
TableHeading	• One or more TableRow child elements • No <TEXT>, <TEXTONLY>, <ANY>, <EMPTY>
TableBody	• One or more TableRow child elements • No <TEXT>, <TEXTONLY>, <ANY>, <EMPTY>
TableFooting	• One or more TableRow child elements • No <TEXT>, <TEXTONLY>, <ANY>, <EMPTY>
TableRow	• One or more TableCell child elements • No <TEXT>, <TEXTONLY>, <ANY>, <EMPTY>
TableCell	• All child elements are allowed, except Table and table parts • <TEXT>, <TEXTONLY>, <ANY>, <EMPTY> allowed

Default GeneralRule

If you import an EDD with an empty GeneralRule element into a document, FrameMaker inserts a default GeneralRule. Here are the default general rules for table parts:

Element Type	Default GeneralRule
Table	TITLE?, HEADING?, BODY, FOOTING?
TableTitle	<ANY>
TableHeading	ROW+
TableBody	ROW+
TableFooting	ROW+
TableRow	CELL+
TableCell	<ANY>

Referencing and Defining the Table Element

Exercise 1: Referencing a Table Element in a GeneralRule

In this exercise, you will redefine the **Section** element's **GeneralRule** to contain optional **Table** elements.

1. If it is not already open, from your class files directory, open `EDD.fm`, the EDD you've been modifying throughout the class.

   If you did not finish the previous chapter's modifications to the EDD, please open **Chapter 4-start Tables.fm** instead, and save it in your class files directory as `EDD.fm`.

For instructions on downloading class files, see "Downloading class files" on page 1.

2. In the EDD, locate the element definition for the **Section** element.
3. In the **Document Window**, change the **GeneralRule** as follows:

```
Head, (Para, (WarnNote | List | Table)?)+, (Section, Section+)?
```

```
[[Element (Container): Section]
    [[General rule: Head, (Para, (WarnNote | List | Table)?)+, (Section, Section+)?]]]
```

4. Save your changes.

Exercise 2: Defining a Table Element

In this exercise, you will define the **Table** element. At a minimum, FrameMaker requires that a table have a body of one row with one cell, but you will put tighter restrictions on your table by creating a general rule requiring a **TableTitle** and a **TableHeading**

In subsequent exercises, you will put even tighter restrictions on your table.

To define the **Table** element:

1. Add an element definition to the EDD and tag it **Table**.
   a. In the **Structure View**, click below the last **Element**.
   b. From the **Elements** panel, insert an **Element**.

      An **Element** and **Tag** child element appear.
   c. In the **Tag** element, type: `Table`

2. Define **Table** as a **Table** with a **GeneralRule** of **TableTitle, TableHeading, TableBody, TableFooting?**
   a. Click below the **Tag** element.
   b. From the **Elements** panel, insert **Table**.
   c. A **Table** element and **GeneralRule** child element appear.
   d. In **GeneralRule** element, type:
   `TableTitle, TableHeading, TableBody, TableFooting?`

[[**Element (Table):** Table]¶ [[**General rule:**〉TableTitle, TableHeading, TableBody, TableFooting?]]]]§

3. Save your changes.

Next you will define the new table parts you referenced in this rule.

Defining Table Part Elements

Exercise 3: Defining a TableTitle Element

To define the **TableTitle** element:

1. Add an element definition to the EDD and tag it **TableTitle**.
   a. In the **Structure View**, click below the last **Element**.
   b. From the **Elements** panel, insert an **Element**.
   An **Element** and a **Tag** child element appear.
   c. In the Tag element, type: `TableTitle`
2. Define **TableTitle** as a **TableTitle** with a **GeneralRule** of <TEXT>
   a. Click below the **Tag** element.
   b. From the **Elements** panel, insert **TableTitle**.
   A **TableTitle** element and a **GeneralRule** child element appear.
   c. In the **GeneralRule** element, type: `<TEXT>`

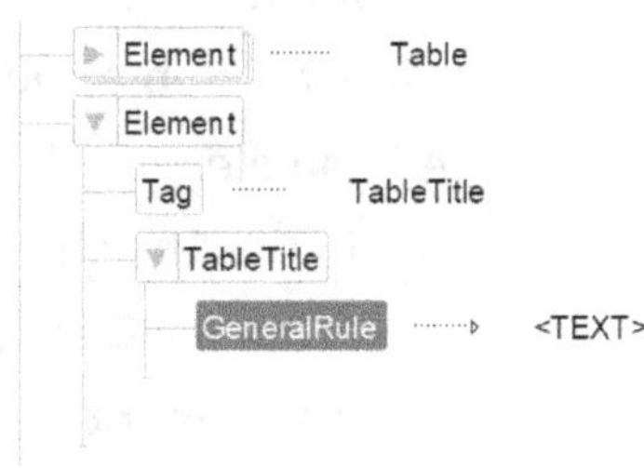

3. Save your changes.

Exercise 4: Defining a TableHeading Element

In this exercise, you will define the **TableHeading** element, by:

- Inserting an **Element** and typing the **Tag** as **TableHeading**
- Inserting a **TableHeading** element to specify **TableHeading** as a table heading
- Typing the **GeneralRule** for the **TableHeading** element's contents: one **TableRow** element

To define the **TableHeading** element:

1. Add an element definition to the EDD and tag it **TableHeading**.
   a. In the **Structure View**, click below the last **Element**.
   b. From the **Elements** panel, insert an **Element**.
      An **Element** and a **Tag** child element appear.
   c. In the **Tag** element, type: `TableHeading`
2. Define **TableHeading** as a **TableHeading** with a **GeneralRule** of **TableRow**.
   a. Click below the **Tag** element.
   b. From the **Elements** panel, insert **TableHeading**.
      TableHeading and a child **GeneralRule** element appear.
   c. In the **GeneralRule** element, type: `TableRow`
3. Save your changes.

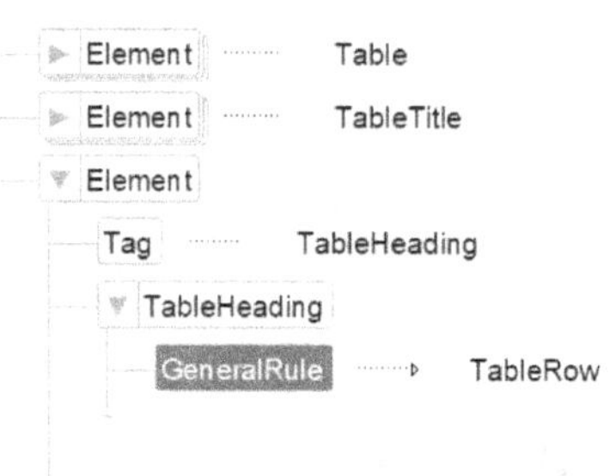

You'll define the **TableRow** later, after defining a few other elements.

Exercise 5: Defining a TableBody Element

In this exercise, you will define the **TableBody** element, by:

- Inserting an **Element** and typing the **Tag** as **TableBody**
- Inserting a **TableBody** element to specify **TableBody** as a table body
- Typing the **GeneralRule** for the **TableBody** element's contents: one or more **TableRow** elements

To define the **TableBody** element:

1. Add an element definition to the EDD and tag it **TableBody**.
   a. In the **Structure View**, click below the last **Element**.
   b. From the **Elements** panel, insert an **Element**.
      An **Element** and a **Tag** child element appear.
   c. In the **Tag** element, type: `TableBody`
2. Define **TableBody** as a **TableBody** with a **GeneralRule** of **TableRow+**
   a. Click below the **Tag** element.
   b. From the **Elements** panel, insert **TableBody**.
      TableBody element and **GeneralRule** child element appear.
   c. In the **GeneralRule** element, type: `TableRow+`
3. Save your changes.

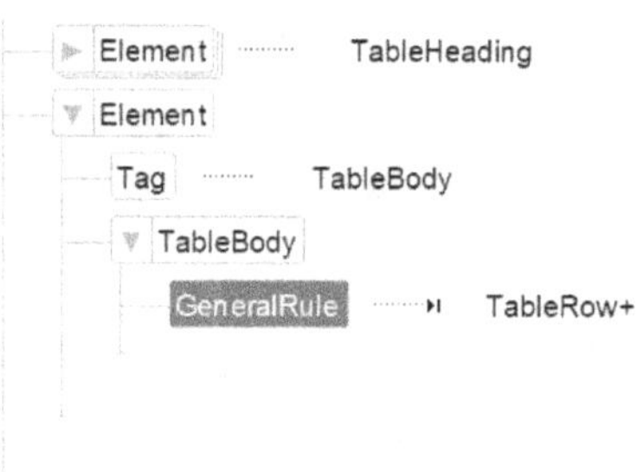

Exercise 6: Defining a TableFooting Element

In this exercise, you will define the **TableFooting** element, by:

- Inserting an **Element** and typing the **Tag** as **TableFooting**
- Inserting a **TableFooting** element to specify **TableFooting** as a table footing
- Typing the **GeneralRule** for the **TableFooting** element's contents: one **TableRow** element

To define the **TableFooting** element:

1. Add an element definition to the EDD and tag it **TableFooting.**
   a. In the **Structure View**, click below the last **Element.**
   b. From the **Elements** panel, insert an **Element.**
      An **Element** and a **Tag** child element appear.
   c. In the **Tag** element, type: `TableFooting`

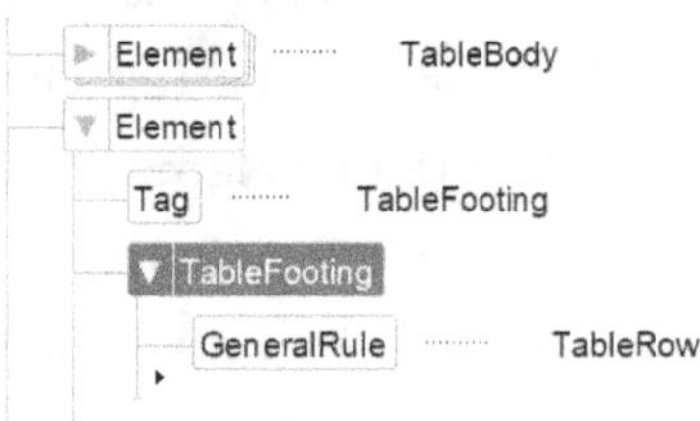

2. Define **TableFooting** as a **TableFooting** with a **GeneralRule** of **TableRow.**
   a. Click below the **Tag** element.
   b. From the **Elements** panel, insert **TableFooting.**
      TableFooting element and **GeneralRule** child element appear.
   c. In the **GeneralRule** element, type: `TableRow`
3. Save your changes.

Exercise 7: Defining a TableRow Element

In this exercise, you will define the **TableRow** element, by:

- Inserting an **Element** and typing the **Tag** as **TableRow**
- Inserting a **TableRow** element to specify **TableRow** as a table row
- Typing the **GeneralRule** for the **TableRow** element's contents: one **Name**, one **Num**, one **Count**

To define the **TableRow** element:

1. Add an element definition to the EDD and tag it **TableRow.**
   a. In the **Structure View**, click below the last **Element.**
   b. From the **Elements** panel, insert an **Element.**
      An **Element** and a child **Tag** child element appear.
   c. In the **Tag** element, type: `TableRow`

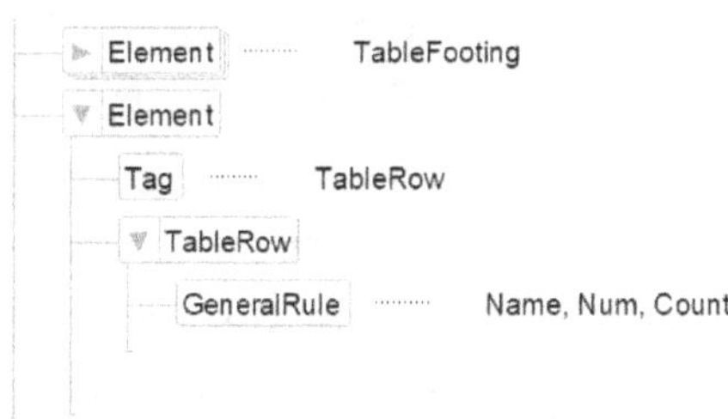

2. Define **TableRow** as a **TableRow** with a **GeneralRule** of **Name, Num, Count.**
   a. Click below the **Tag** element.
   b. From the **Elements** panel, insert **TableRow.**
      A **TableRow** element and **GeneralRule** child element appear.
   c. In the **GeneralRule** element, type: `Name, Num, Count`

3. Save your changes.

Exercise 8: Defining TableCell Elements

In this exercise, you will define separate Name, Num, and Count table cell elements, with a general rule of <TEXT>

To define the **TableCell** elements:

1. Add an element definition to the EDD and tag it **Name**.
   a. In the **Structure View**, click below the last **Element**.
   b. From the **Elements** panel, insert an **Element**.

      An **Element** and a **Tag** child element appear.
   c. In the Tag element, type: `Name`
2. Define **Name** as a **TableCell** with a **GeneralRule** of <TEXT>
   a. Click below the **Tag** element.
   b. From the **Elements** panel, insert **TableCell**.

      A **TableCell** element and a **GeneralRule** child element appear.
   c. In the **GeneralRule** element, type: `<TEXT>`

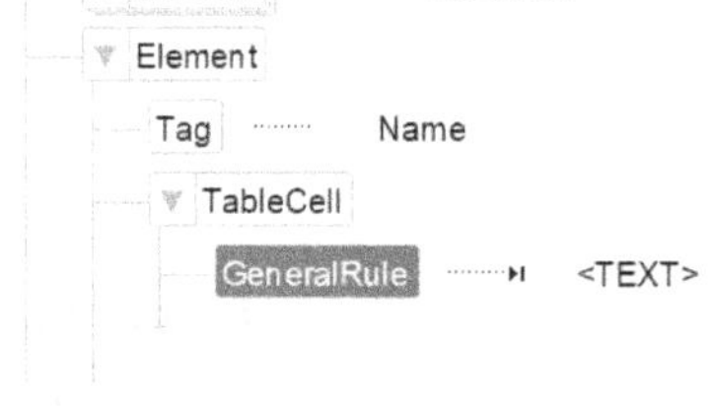

3. Add an element definition to the EDD and tag it **Num**.
   a. In the **Structure View**, click below the last **Element**.
   b. From the **Elements** panel, insert an **Element**.

      An **Element** and a **Tag** child element appear.
   c. In the **Tag** element, type: `Num`

Consider using **Copy/Paste** for duplicating similar structure

4. Define **Num** as a **TableCell** with a **GeneralRule** of <TEXT>
   a. Click below the **Tag** element.
   b. From the **Elements** panel, insert **TableCell**.

      A **TableCell** element and a **GeneralRule** child element appear.
   c. In the **GeneralRule** element, type: `<TEXT>`

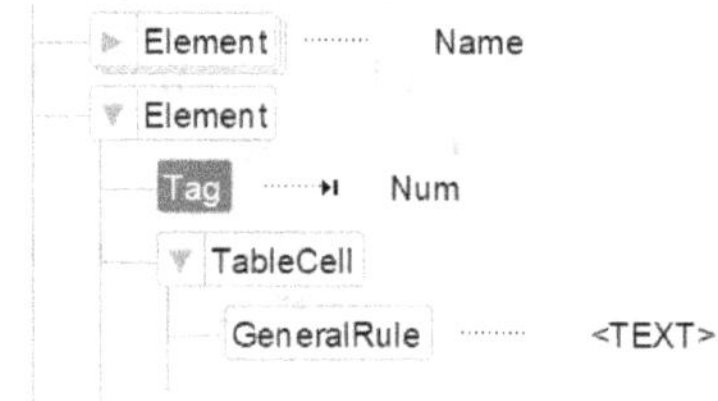

5. Add an element definition to the EDD and tag it **Count**.
   a. In the **Structure View**, click below the last **Element**.
   b. From the **Elements** panel, insert an **Element**.

      An **Element** and a **Tag** child element appear.
   c. In the **Tag** element, type: `Count`

6. Define **Count** as a **TableCell** with a **GeneralRule** of <TEXT>
    a. Click below the **Tag** element.
    b. From the **Elements** panel, insert **TableCell**.

       A **TableCell** element and a **GeneralRule** child element appear.
    c. In the **GeneralRule** element, type: `<TEXT>`

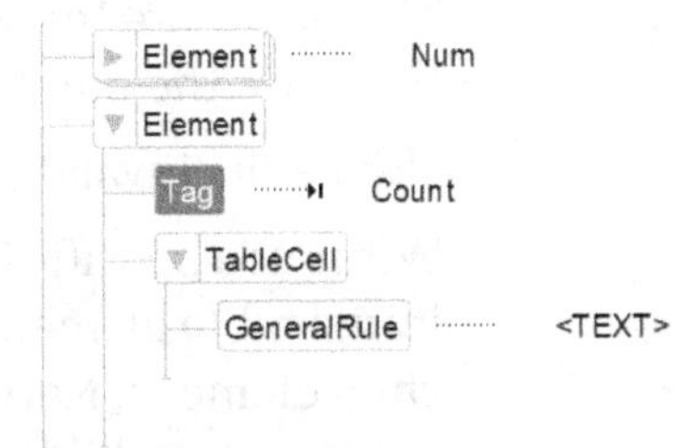

7. Save your changes.

Reimporting and Testing

Exercise 9: Reimporting and Testing the Table

In this exercise, you will reimport the EDD into the structured template and test your element definitions for Section, Table, and child elements of Table.

1. Reimport the element definitions from your EDD into your test document. If you get errors when importing the element definitions, correct your EDD and reimport.

   Continue to the next step when your import occurs without errors.
2. In the **Structure View**, click on the line descending from any **Section** element but after a **Para** element.
3. From the **Elements** panel, insert a **Table** element.

   The **Insert Table** dialog appears.
4. In the **Insert Table** dialog, specify:
   - Columns: 3
   - Body Rows: 3
   - Heading Rows: 1
   - Footing Rows: 1

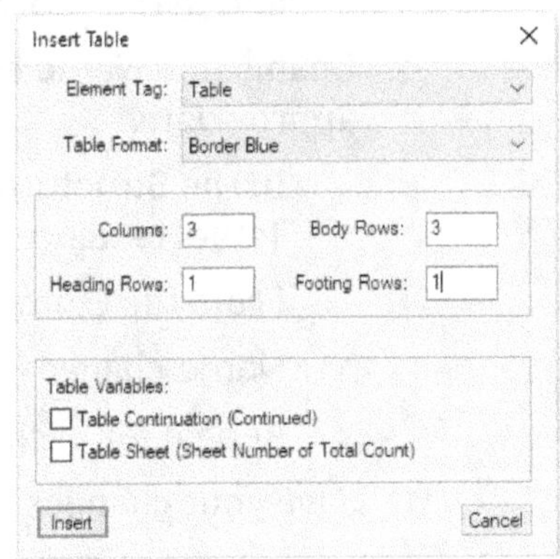

5. Click **Insert**.

   The **Document Window** shows the table.

[[section head]
[para]

Table 1:

[[section head]]
[para]]

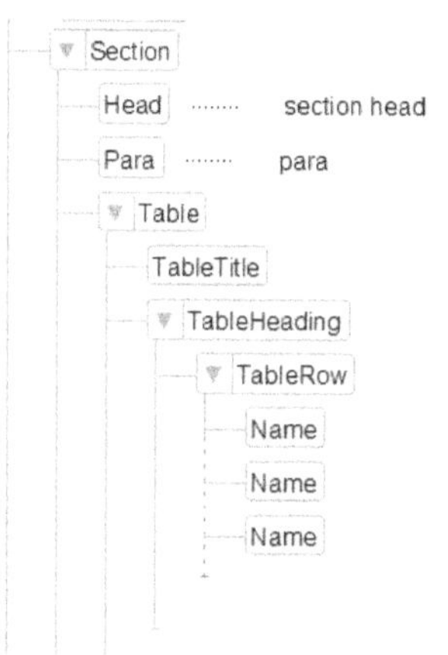

The **Structure View** shows the table's overall structure, based on the number of rows and columns you specified.

Notice the invalid structure of each **TableRow** element.

Without a specific **InitialStructurePattern** rule to define initial table elements, by default, FrameMaker uses the first child element (**Name**) in the **GeneralRule** for **TableRow** and repeats that child element to create all the columns you specify in the **Insert Table** dialog.

Later, you will change this behavior by specifying an **InitialStructurePattern** for **TableRow**. In the meantime, you will change the **TableRow's** child elements into the correct child elements.

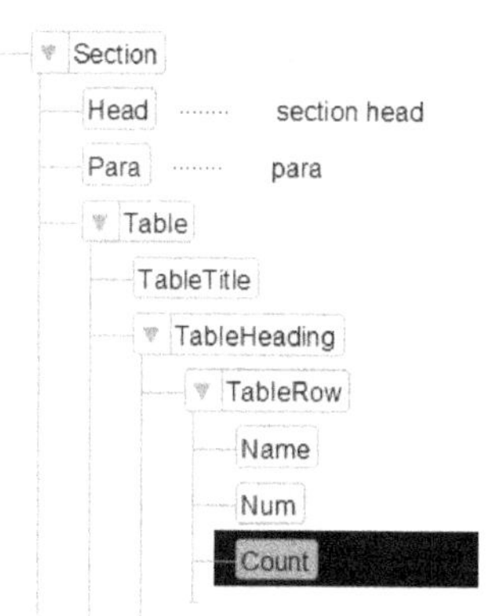

6. Change the **Name** elements to **Num** elements where appropriate.
    a. In the **Structure View**, select the second **Name** element in the first **TableRow** in the **TableHeading**.
    b. From the **Elements** panel, select **Num** and click **Change.**

       Name changes to **Num**.
    c. Repeat for each additional **TableRow** in the **TableBody** and in the **TableFooting**.
7. Change the **Name** elements to **Count** elements where appropriate.
    a. In the **Structure View**, select the third **Name** element in the first **TableRow** in the **TableHeading**.
    b. From the **Elements** panel, select **Count** and click **Change.**

       Name changes to **Count**.
    c. Repeat for each additional **TableRow** in the **TableBody** and in the **TableFooting**.
8. Save your changes.

Exercise 10: Testing the Parts of the Table

In this exercise, you will continue testing the parts of the **Table—TableTitle, TableBody, TableHeading, TableFooting, TableRow, Name, Num, Count.**

1. In the **Structure View**, click to the right of the **TableTitle** element and type:

   ```
   Items needing routine maintenance
   ```

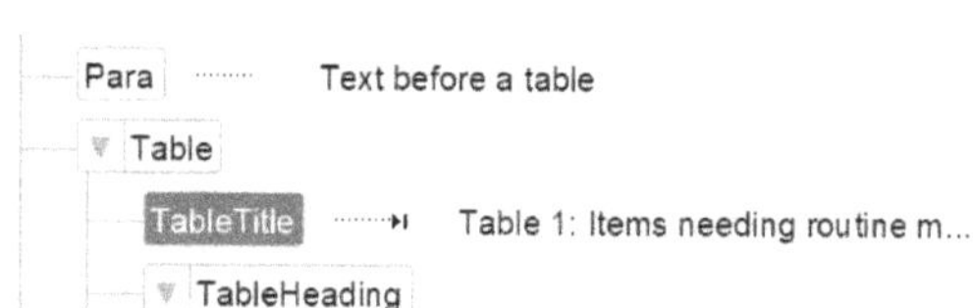

2. Click in at least one of each—**Name, Num,** and **Count**— and enter text. (not shown)

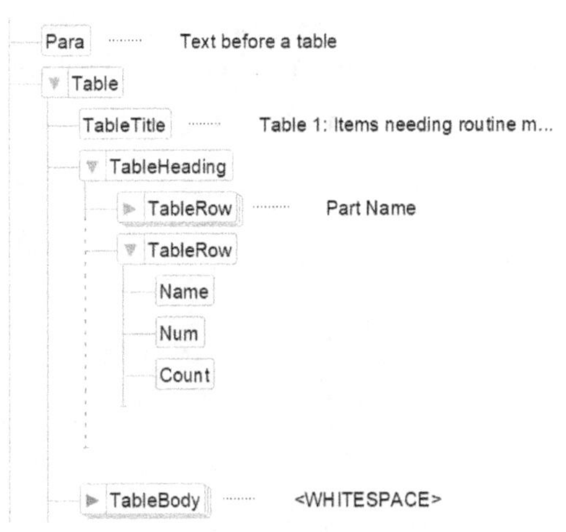

3. Add another row to the **TableHeading** in the **Document View.**
   a. Click in the heading row of the table.
   b. Press **Control-Return.**

      A second TableRow is added to TableHeading, but only one is allowed so the structure shows as invalid.
4. From the **Edit** menu, choose **Undo.**

   The invalid **TableRow** disappears.
5. Delete the **TableHeading.**
   a. In the **Structure View,** select the **TableHeading** element.
   b. Press **Delete.**

      The Clear Table Cells dialog appears.

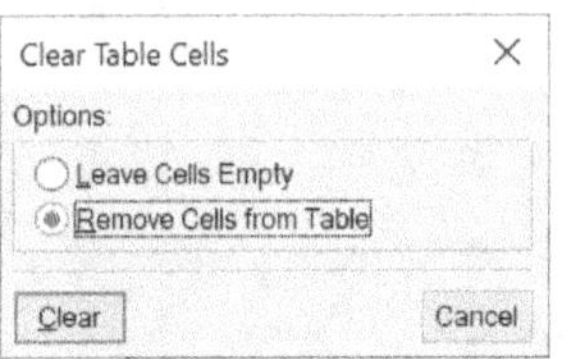

   c. Turn on **Remove Cells from Table.**
   d. Click **Clear.**

      The table heading is deleted.

      A TableHeading is required, so the structure is invalid.

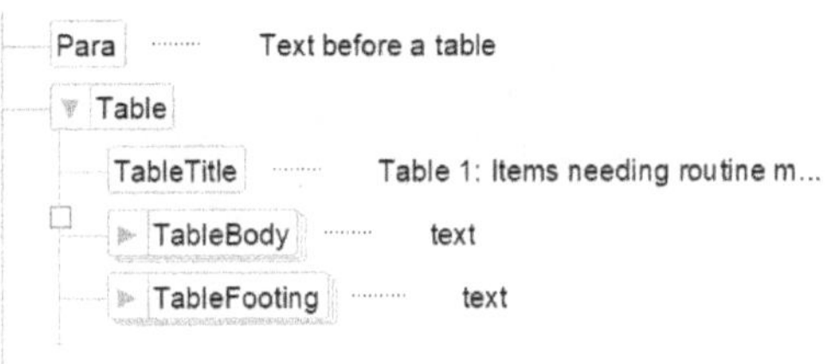

6. From the Edit menu, choose **Undo.**
7. Add another row to the TableFooting.
   a. Click in the footing row of the table.
   b. Press **Control-Return.**

      A second **TableRow** is added to **TableFooting,** but as with the TableHeading, only one is allowed so the structure is invalid.
8. From the **Edit** menu, choose **Undo.**

   The invalid **TableRow** disappears.
9. Delete the **TableFooting** element.
   a. In the **Structure View,** select the **TableFooting** element.
   b. Press **Delete.**

      The **Clear Table Cells** dialog appears.
   c. Turn on **Remove Cells from Table.**
   d. Click **Clear.**

The **TableFooting** disappears, and the structure is valid because the **TableFooting** is optional.

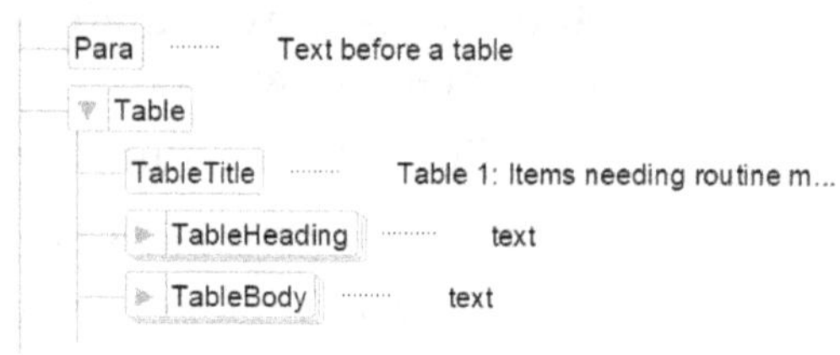

10. Save your changes.

Chapter 5: Tables—InitialStructurePattern and InitialTableFormat

Introduction

This chapter focuses on the **InitialStructurePattern** and **InitialTableFormat** for tables.

Objectives

- Specify **InitialStructurePattern** for **Table** and table part elements
- Specify **InitialTableFormat** for **Table** element
- Reimport and test **InitialStructurePattern** and **InitialTableFormat**

Overview of InitialStructurePattern

InitialStructurePattern is an optional part of element definitions for the following elements:

- Table
- TableHeading
- TableBody
- TableFooting
- TableRow

All tables have these common characteristics:

- Tables must contain a table body
- Table heading, body, or footing elements must have at least one row
- Table rows must have same number of cells as columns in table

When you insert a Table element, you use the **Insert Table** dialog to specify number of rows in heading, body, footing and number of columns. FrameMaker automatically inserts the necessary child elements to build a basic structure.

If you do not specify initial structure, FrameMaker:

- Uses a default **GeneralRule** to give new table or table part its default initial structure
- Builds default initial structure by taking first of each type of table part in table's **GeneralRule**

An **InitialStructurePattern** element can specify:

- The **TableTitle, TableHeading, TableBody, TableFooting** elements that will initially appear in the table element
- The **TableRow** type elements that will appear in the **TableHeading, TableBody, TableFooting** type elements
- The **TableCell** type elements that will appear in the **TableRow** type elements

Specifying an InitialStructurePattern

Exercise 1: Specifying the InitialStructurePattern for a TableRow Element

In this exercise, you will define an **InitialStructurePattern** within your **TableRow** element definition that will insert **Name, Num,** and **Count** elements instead of inserting all **Name** elements.

1. If it is not already open, from your class files directory, open the **EDD.fm** file, the EDD you've been modifying throughout the lessons.

   If you did not finish the previous chapter's modifications to the EDD, please open **Chapter 5-start Initial Table Format.fm** instead, and save it in your class files directory as **EDD.fm**.

For instructions on downloading class files, see "Downloading class files" on page 1.

2. In the EDD, locate the **TableRow** element definition.
3. In the **Structure View**, click below **TableRow**'s **GeneralRule** element.
4. From the **Elements** panel, insert **InitialStructurePattern.**

   An **InitialStructurePattern** element appears.

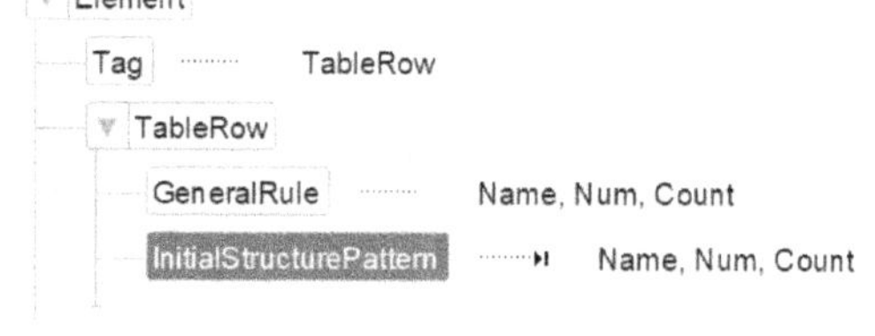

5. In the **InitialStructurePattern** element, type: `Name, Num, Count`
6. Save your changes.

InitialTableFormat—Overview

InitialTableFormat is an optional part of a **Table** element definition.

- It identifies which table format is preselected in the **Insert Table** dialog when user inserts **Table** element
- If no **InitialTableFormat** specified, a **Table** element uses **Format A** as a default format
- Makes sure your table format has a **GeneralRule** that is consistent with **Table** element requirements

If a **Table** element's **GeneralRule** does not specify a **TableTitle** element, the table format should not have one.

InitialTableFormat is only a suggestion—users can change to another format without creating format override

You can specify a table format with:

- **AllContextsRule**—wherever the element appears

- **ContextRule**—when element is in certain context
- **LevelRule**—when element is nested a specified number of levels in a specified ancestor

You'll learn more about context and level rules later in this course.

Specifying an InitialTableFormat

Exercise 2: Specifying an InitialTableFormat with an AllContextsRule

In this exercise, you will specify which table format will be preselected in the **Insert Table** dialog when the user inserts a **Table** element.

As noted, the user can still select a different table format without incurring a format rule override.

1. In the EDD, locate the **Table** element definition.
2. In the **Structure View**, click below **Table**'s **GeneralRule** element.
3. From the **Elements** panel, insert **InitialTableFormat**.

   An **InitialTableFormat** element appears.

At this point, you could specify a context-specific or level-specific rule. Since these rules are covered in the next chapter, in this exercise you will insert the simpler **AllContextsRule**.

4. Insert an **AllContextsRule** element.

   An **AllContextsRule** element appears.
5. Insert a **TableFormat** element.

   A **TableFormat** element appears.
6. In the **TableFormat** element, type:
   `Format B`
7. Save your changes.

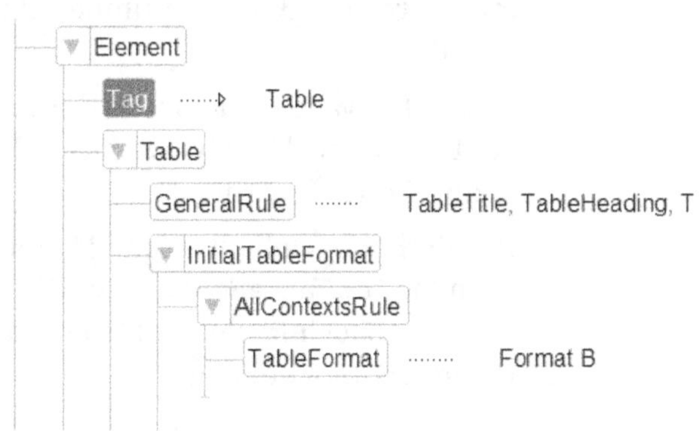

Reimporting and Testing

Exercise 3: Reimporting/Testing InitialStructurePattern and InitialTableFormat

In this exercise, you will reimport your modified EDD into your test document and test your **InitialStructurePattern** on **TableRow** and **InitialTableFormat** on **Table**.

1. Reimport the element definitions from your EDD into your test document. If you get errors when importing the element definitions, correct your EDD and reimport.

   Continue to the next step when your import occurs without errors.
2. In the **Structure View**, click below a **List** element.

3. From the **Elements** panel, insert a **Para** element.
4. Click below the **Para** element.
5. Insert a **Table** element.

   The **Insert Table** dialog appears, with **Format B** preselected as you defined in the EDD.

6. In the **Insert Table** dialog, specify:
   - Columns: 3
   - Body Rows: 3
   - Heading Rows: 1
   - Footing Rows: 1
7. Click **Insert.**

   The **Document Window** displays the table.

Table 1: §

§	§	§
§	§	§
§	§	§
§	§	§
§	§	§

The **Structure View** displays the table's overall structure, based on the number of rows and columns you specified.

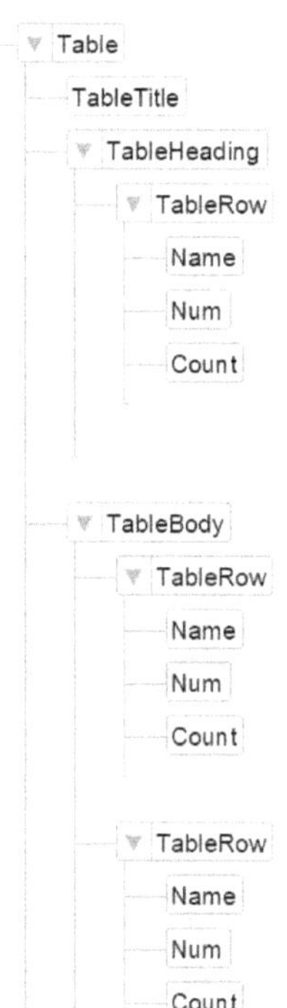

Notice the valid structure of the **TableRow** elements which now contain **Name, Num, Count** as defined in the EDD, rather than three **Name** elements.

Recall that by default, FrameMaker uses the first child element found in the **TableRow GeneralRule** (in this case, **Name**) and repeats that child element to create all the columns you specify in the **Insert Table** dialog.

You changed this behavior by specifying an **InitialStructurePattern** for **TableRow**.

8. Save your changes.

Chapter 6: Inclusions and Exclusions

Introduction

Inclusions allow elements to occur anywhere inside a defined element or its descendants. They are useful for infrequently used elements that might be necessary in multiple places within the hierarchy, such as **Footnote** or **Term**.

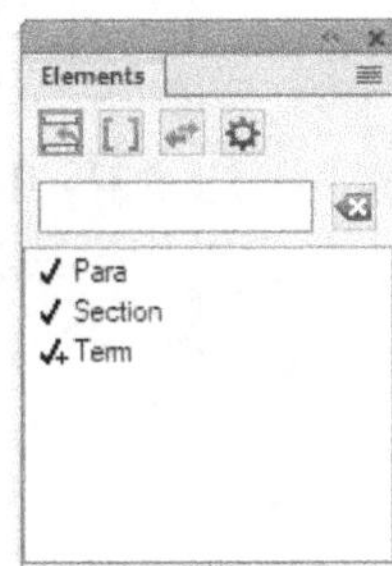

Inclusions greatly simplify the general rules for elements. They also help your audience manage their **Elements** panel view by providing them with the option to list less-often used elements below a list of more-often used valid elements via the **Elements** panel **Options** (⚙).

In the example shown here, when inclusions are set to display after other valid elements, the **Term** element in the screen capture to the right displays the inclusion icon (✓)and is listed after other valid elements.

This chapter focuses on defining **Inclusions** and **Exclusions** for **Container**, **Footnote**, **Table** and table part elements.

Objectives

- Specify use of an inclusion
- Define the included element
- Specify an exclusion of an element
- Reimport and test inclusions and exclusions

Defining Inclusions

Inclusions are an optional part of element definitions for:

- Containers
- Table, TableTitle, TableHeading, TableBody, TableFooting, TableRow, TableCell
- Footnotes

Exercise 1: Specifying Inclusions

In this exercise, you will define a new **Container** element tagged **Term** to contain the names of other documents mentioned throughout the chapters of the maintenance manuals. Rather than referencing **Term** in the **GeneralRule** for each element, you will define it as an inclusion for **Section**. **Term** can then be included anywhere within **Section** and its descendants.

1. If it is not already open, from your class files directory, open **EDD.fm**, the EDD you've been modifying throughout the class.

   If you did not finish the previous chapter's modifications to the EDD, please open **Chapter 6-start Inclusions Exclusions.fm** instead, and save it in your class files directory as **EDD.fm**.

For instructions on downloading class files, see "Downloading class files" on page 1.

2. In the EDD, create a new **Element** definition and tag it **Term**.
    a. In the **Structure View**, click below the last **Element** definition.
    b. From the **Elements** panel, insert a new **Element.**

       An **Element** and **Tag** child element appear.
    c. In the **Tag** element, type: **Term**

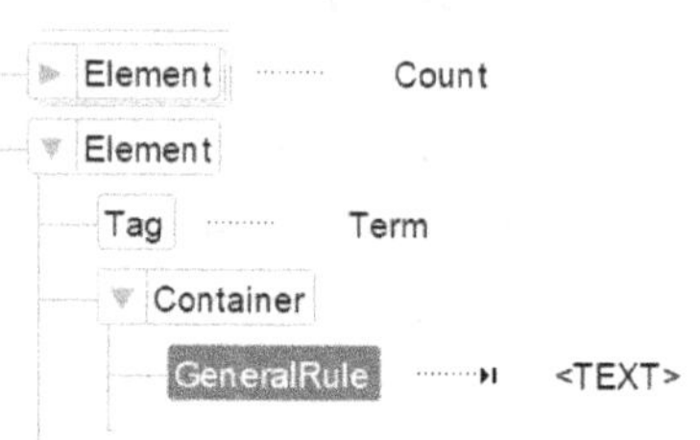

3. Define **Term** as a **Container** with a **GeneralRule** of <TEXT>
    a. Click below the **Tag** element.
    b. From the **Elements** panel, insert **Container.**

       A **Container** element and **GeneralRule** child element appear.
    c. In the **GeneralRule** element, type: `<TEXT>`
4. In the **Structure View**, locate the **Section** element.

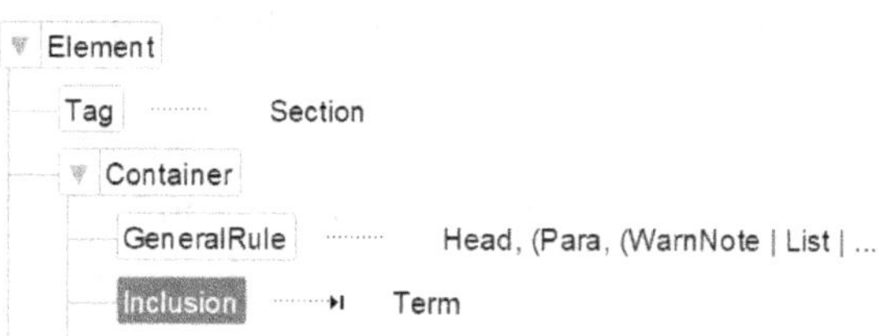

5. Click below the **GeneralRule** of the **Section** element.
6. From the **Elements** panel, insert **Inclusion.**

   An **Inclusion** element appears.
7. In the **Inclusion** element, type: `Term`
8. Save your changes.

You'll test this change after also defining an exclusion.

Defining Exclusions

Exclusions are an optional part of element definitions for the following types of elements:

- **Container**
- **Table, TableTitle, TableHeading, TableBody, TableFooting, TableRow, TableCell**
- **Footnote**

Exclusions specify:

- Elements that cannot occur anywhere in a defined element or its descendants
- Exclusions are often used to negate inclusions

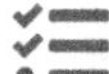

Exercise 2: Specifying Exclusions

In this exercise, you will exclude **Term** from **Head** and **Table** so that it cannot appear within any **Head** element or any **Table** element or its descendant parts.

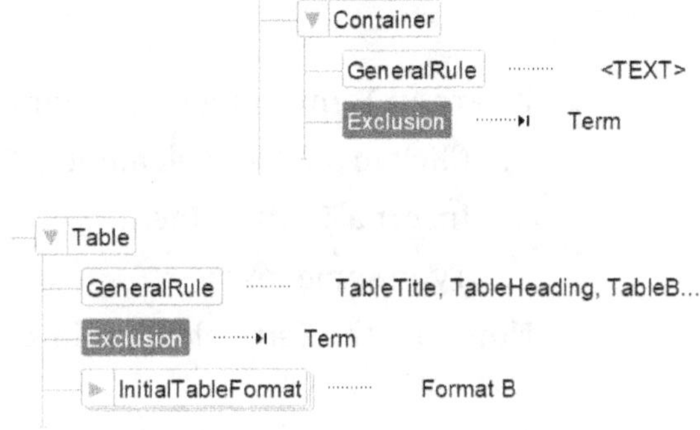

1. In the **Structure View**, locate the **Head** element definition.
2. Click below the **GeneralRule** for the **Head** element.
3. From the **Elements** panel, insert **Exclusion**.

   An **Exclusion** element appears.
4. In the Exclusion element, type: `Term`
5. In the **Structure View**, locate the **Table** element.
6. Click below the **Table's GeneralRule** element.
7. From the **Elements** panel, insert **Exclusion**.

   An **Exclusion** element appears.
8. In the **Exclusion** element, type: `Term`
9. Save your changes.

Reimporting and Testing Inclusions and Exclusions

Exercise 3: Reimporting and Testing Inclusions and Exclusions

In this exercise, you will reimport the EDD into the structured template and test your **Term** element and inclusions and exclusions.

1. Reimport the element definitions from your EDD into your test document. If you get errors when importing the element definitions, correct your EDD and reimport.

   Continue to the next step when your import occurs without errors.

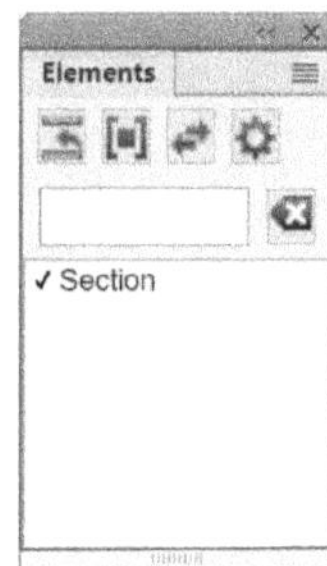

2. In the **Structure View**, click anywhere on the line descending from the **Chapter** element.

   The **Elements** panel does not show **Term** as an inclusion, because you defined it as an inclusion on **Section**, not **Chapter**.
3. Click in the **Title** element.

   The **Elements** panel does not show **Term** as an inclusion, because you defined it as an inclusion on **Section**, and **Title** is a child of **Chapter**, not **Section**.

4. Click anywhere on the line descending from the **Section** element.

   The **Elements** panel shows **Term** as an inclusion. Your listed elements may differ depending on where you position your cursor in the structure view.

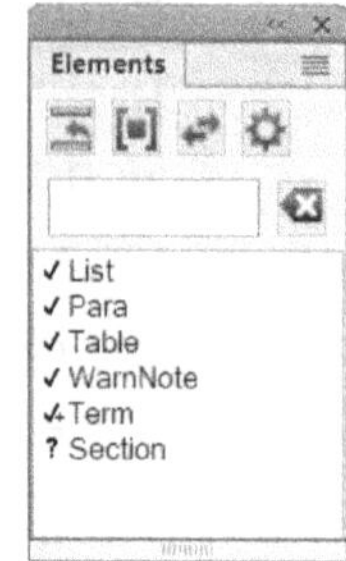

5. From the **Elements** panel, insert an **Term** element and insert some text.

6. Insert an **Term** element in a Para element.

   a. Click in any **Para** element.

   b. Insert a **Term** element.

   c. Type some text.

   Note that the **Term** element forces a line break in the **Para** element.

   Later, you will define the **Term** element's formatting as a text range so that the break will disappear.

7. Click in any **Head** element.

   Although **Head** is a child of **Section**, the **Elements** panel does not show **Term** because you excluded **Term** from **Head.** (not shown)

8. Click in any **Num** table cell element (in the table, not shown).

   Although **Num** is a descendant of **Section**, the **Elements** panel does not show **Term**, because you excluded **Term** from **Table** (and descendants of **Table**, like **Num**).

9. Save your changes.

Chapter 7: AutoInsertions

Introduction

This chapter focuses on defining element autoinsertions for container elements.

Objectives

- Specify autoinserted child elements
- Specify autoinserted nested child elements
- Reimport and test the AutoInsertions

Overview of Autoinsertions

Autoinsertions are an optional part of element definitions for things like containers and table cells.

- They identify a child element to be inserted automatically with its parent
- Autoinsertions can also identify nested children ("grandchild" then "great-grandchild," etc.) to insert automatically
- Autoinsertions cannot automatically insert sibling elements

FrameMaker opens dialogs and windows as needed to specify:

- Attributes
- Table number of rows and columns
- Graphic positioning
- Cross-reference source
- Variable selection
- Marker text

Specifying Autoinserted Child Elements

Exercise 1: Specifying Autoinserted Title Element

In this exercise, you will specify that the **Title** child element will appear automatically when you insert a **Chapter** element. When a first child element is required, specifying it as an **AutoInsertion** speeds up the authoring process.

1. If it is not already open, from your class files directory, open **EDD.fm**, the EDD you've been modifying throughout the class.

   If you did not finish the previous chapter's modifications to the EDD, please open **Chapter 7-start AutoInsertions.fm** instead, and save it in your class files directory as **EDD.fm**.

For instructions on downloading class files, see "Downloading class files" on page 1.

2. In the EDD, locate the **Chapter** element definition.
3. Click below the **GeneralRule** element for **Chapter** (or click below the **ValidHighestLevel** element if you inserted it below, rather than above, the **GeneralRule** for **Chapter**).
4. From the **Elements** panel, insert an **AutoInsertions** element.

   An **AutoInsertions** element and an **InsertChild** element appear.
5. In the **InsertChild** element, type: `Title`
6. Save your changes.

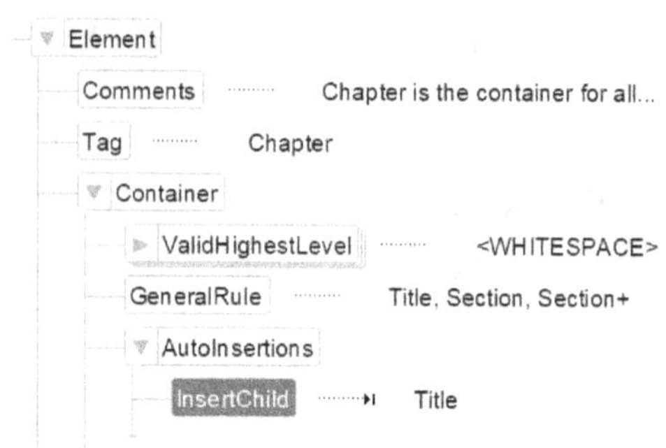

Exercise 2: Specifying Autoinserted Head Element

In this exercise, you will specify that the **Head** child element will appear automatically when you insert the **Section** element.

1. In the EDD, locate the **Section** element definition.
2. Click below the **Section**'s **Inclusion** element.
3. From the **Elements** panel, insert **AutoInsertions.**

   An **AutoInsertions** element with **InsertChild** element appear.
4. In the **InsertChild** element, type: `Head`
5. Save your changes.

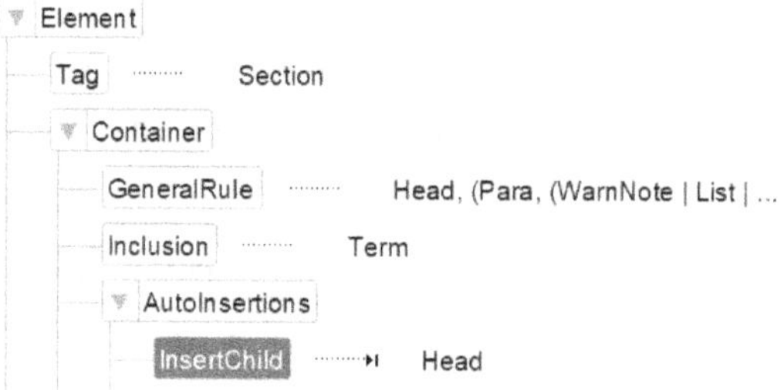

Exercise 3: Reimporting and Testing Autoinserted Child Elements

In this exercise, you will reimport the EDD into the structured template, verify that **Allow Automatic Insertion of Children** is turned on, and test your **AutoInsertions** on **Chapter** and **Section.**

1. Reimport the element definitions from your EDD into your test document. If you get errors when importing the element definitions, correct your EDD and reimport.

   Continue to the next step when your import occurs without errors.
2. Verify that **Allow Automatic Insertion of Children** is turned on.
   a. Select **Element>New Element Options.**

      The **New Element Options** dialog appears.
   b. If not already on, turn on **Allow Automatic Insertion of Children.**
   c. Click **Set.**

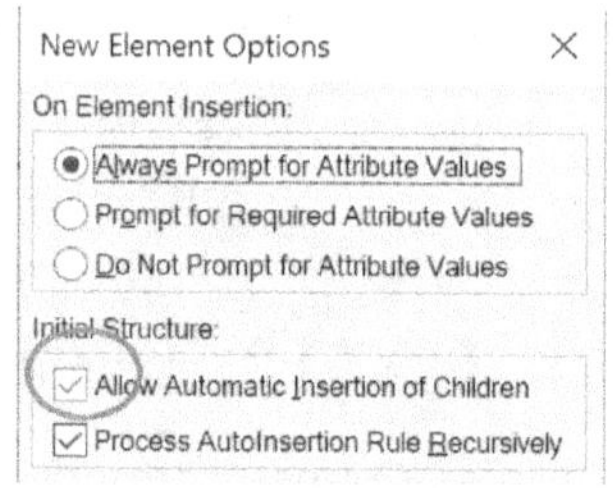

Although **AutoInsertions** are defined in the EDD, the feature can be disabled in the document via this dialog.

3. In the **Structure View**, select and delete the **Chapter** element.

Since **Chapter** is the highest-level element, all your test content disappears. (Yes, this is what you need to do!)

4. From the **Elements** panel, insert a **Chapter** element.

   The **Chapter** element appears, and is followed by a **Title** child element, courtesy of the **InsertChild** element in your EDD.

5. Type sample content into the **Title** element.
6. Click below the **Title** element.
7. Insert a **Section** element.

   A **Section** element appears, along with a child element of **Head**.

8. Type sample content into the **Head** element.
9. Save your changes.

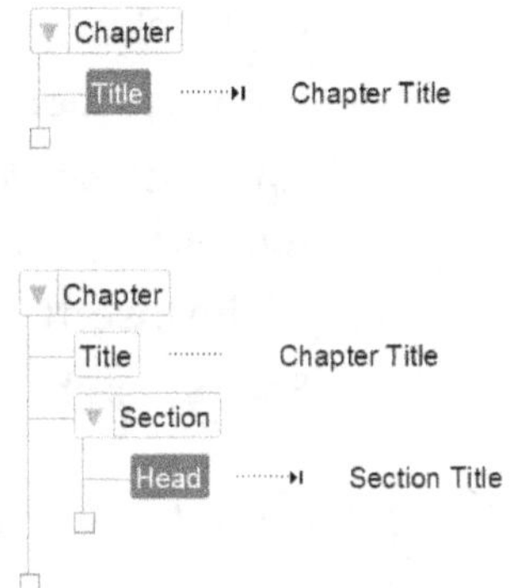

Specifying Autoinserted Child and Nested Child Elements

Exercise 4: Specifying Autoinserted Item and Nested Para Elements

In this exercise, you will redefine the **Item** element to contain one or more **Para** elements with optional **WarnNote** elements, rather than just <TEXT>. (In a later exercise, you will define formatting so that “first” and “not first” **Para** elements in **Item** elements have different formatting.)

After redefining **Item**, you will specify **AutoInsertions** so that:

- Lists will automatically insert **Item** child and **Para** nested child
- Items will automatically insert **Para** child

1. In the EDD, locate the **Item** element definition.
2. In the **Document Window**, replace the **GeneralRule** as follows:

```
(Para, WarnNote?)+
```

With this change, rather than allowing only a single paragraph of text (<TEXT>), an Item can now have one or more Para elements, along with optional WarnNote elements between paragraphs.

3. In the **Structure View**, locate the **List** element definition.
4. Click below the **List's GeneralRule** element.
5. From the **Elements** panel, insert **AutoInsertions**.

   AutoInsertions element with **InsertChild** element appear.

6. In the **InsertChild** element, type: `Item`
7. Click below the **InsertChild** element.
8. From the **Elements** panel, insert **InsertNestedChild**.

   An **InsertNestedChild** element appears.

9. In the **InsertNestedChild** element, type: `Para`

   Now, when you insert a **List** element, an **Item** child element and a nested **Para** child element will appear.

   However, this does not mean that when you insert an additional **Item** element, its **Para** child element will appear. To acheive this you will define an **Item** autoInsertion separately.

10. In the **Structure View**, locate the **Item** element definition.
11. Click below the **Item's GeneralRule** element.
12. From the **Elements** panel, insert **AutoInsertions.**

    An **AutoInsertions** element with **InsertChild** element appears.

13. In the **InsertChild** element, type: `Para`
14. Save your changes.

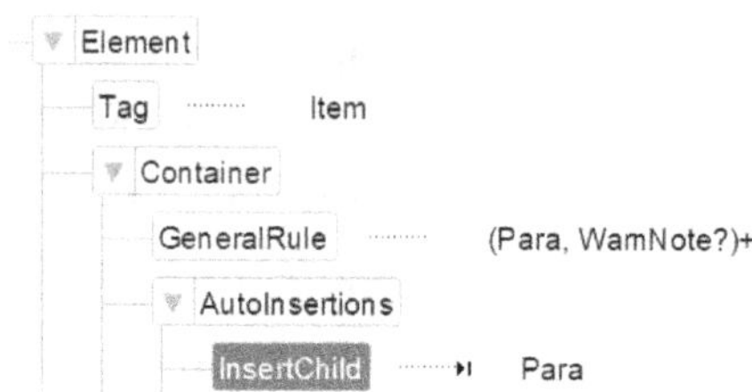

Reimporting and Testing Item and Para Autoinsertions

Exercise 5: Reimporting and Testing Autoinserted Nested Child Elements

In this exercise, you will reimport the EDD into the structured template and test your autoinsertions on **List** and **Item.**

1. Reimport the element definitions from your EDD into your test document. If you get errors when importing the element definitions, correct your EDD and reimport.

   Continue to the next step when your import occurs without errors.

2. In the **Structure View**, click below the **Head** element.
3. From the **Elements** panel, insert a **Para** element.
4. Click below the **Para** element.
5. Insert a **List** element.

   List element and **Item** child element and **Para** nested child element appear.

6. Click below the **Item** element, on the line descending from the **List** element.
7. Insert an **Item** element.

   Item element and **Para** child element appear.

8. Save your changes.

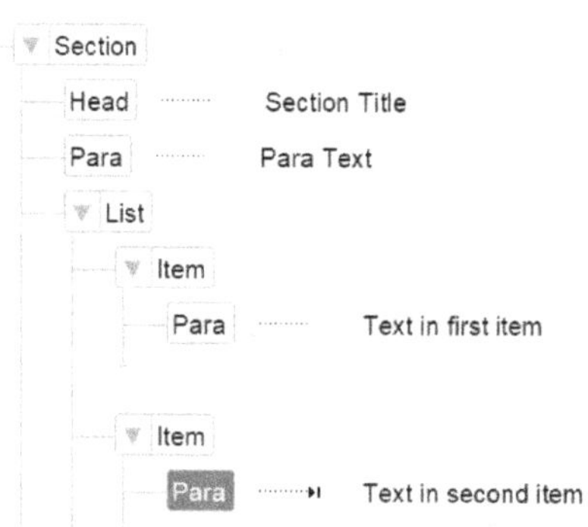

Chapter 8: Defining and Formatting Objects

Introduction

This chapter focuses on defining objects like cross-references, equations, graphics, and markers. These types of elements usually require formatting, using things like the **InitialObjectFormat** and the **SystemVariableFormatRule** for system variables.

Objectives

- Define a **CrossReference** element (labeled **XRef**) and specify its **InitialObjectFormat**, referring to a cross-reference format stored in a document
- Define an **Equation** element (labeled **EQ**)and specify its **InitialObjectFormat**, referring to the equation's size
- Define a **Figure** element, which will contain both **Caption** and **Graphic** elements
- Define a **Graphic** element (labeled **Graphic**)and specify its **InitialObjectFormat**, using the **Import File** or **Anchored Frame** dialog
- Define a **Marker** element (labeled **IndexEntry**)and specify its **InitialObjectFormat**, referring to the marker type
- Define a **SystemVariable** element (labeled **Date**)and specify its **SystemVariableFormatRule**, referring to a variable name
- Reimport and test object elements and formats

Overview of InitialObjectFormat

The **InitialObjectFormat** is an optional part of an object's element definition and specifies the default formatting applied when inserting an object.

The **InitialObjectFormat** is only a suggestion—an author can choose another format without creating format override.

Object	InitialObjectFormat
CrossReference	Name of cross-reference format stored in the documents to be preselected in Cross-Reference dialog when user inserts element
Equation	Equation size of Small, Medium or Large
Graphic	ImportedGraphicFile or AnchoredFrame
Marker	Marker type to be preselected in Insert Marker dialog when user inserts element

If an **InitialObjectFormat** is not specified, FrameMaker applies defaults as follows:

Object	InitialObjectFormat
CrossReference	Last Cross-Reference format chosen in dialog
Equation	Medium
Graphic	AnchoredFrame
Marker	Last marker type chosen in Insert Marker dialog or Marker window

There are a number of ways to specify an **InitialObjectFormat**. You can specify a format with:

- **AllContextsRule**—wherever the element appears
- **ContextRule**—when element is in certain context
- **LevelRule**—when element is nested a specified number of levels in a specified ancestor

CrossReference Elements

Exercise 1: Defining a CrossReference and InitialObjectFormat

In this exercise, you will:

- Redefine the **Para** element to contain optional **XRef** elements
- Define the **XRef** element as a **CrossReference**, using the **InitialObjectFormat** of **ElemNumTextPage**

The **InitialObjectFormat** for the **XRef** means that the **ElemNumTextPage** will be preselected in the **Cross-Reference** dialog when the user inserts the **XRef** elements. As with table formatting, the user can select a different cross-reference format without incurring a format rule override.

1. If it is not already open, from your class files directory, open **EDD.fm**, the EDD you've been modifying throughout the class.

   If you did not finish the previous chapter's modifications to the EDD, please open **Chapter 8-start Define Format Objects.fm** instead, and save it in your class files directory as **EDD.fm**.

For instructions on downloading class files, see "Downloading class files" on page 1.

2. In the EDD, locate the **Para** element definition.
3. In the **Document Window**, change the **GeneralRule** as follows:

   (<TEXT> | Footnote **| XRef**)+

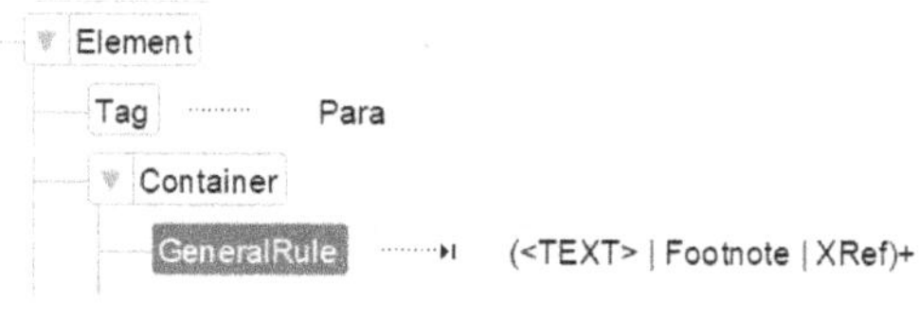

4. Create a new **Element** and tag it **XRef**.
   a. In the **Structure View**, click below the last **Element** definition.
   b. From the **Elements** panel, create a new **Element** definition.

      An **Element** and a **Tag** child element appear.
   c. In the **Tag** element, type: **XRef**
5. Define **XRef** as a **CrossReference**.
   a. Click below the **Tag** element.
   b. From the **Elements** panel, insert a **CrossReference**.

      A **CrossReference** element appears.

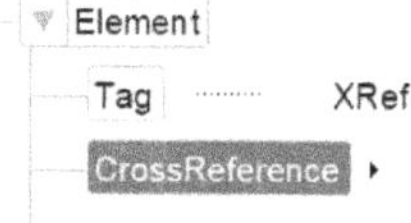

6. From the **Elements** panel, insert an **InitialObjectFormat** element.

   An **InitialObjectFormat** element appears.

7. Insertan **AllContextsRule.**

   An **AllContextsRule** element appears.

8. Inserta **CrossReferenceFormat.**

   A **CrossReferenceFormat** element appears.

Element
Tag XRef
CrossReference
InitialObjectFormat
AllContextsRule
CrossReferenceFormat ElemNumTextPage

9. In the **CrossReferenceFormat** element, type: `ElemNumTextPage`
10. Save your changes.

Exercise 2: Reimporting and Testing the Cross-Reference Element

In this exercise, you will reimport the EDD into the structured template and test your element definitions for the **Para** and **XRef** elements.

1. Reimport the element definitions from your EDD into your test document. If you get errors when importing the element definitions, correct your EDD and reimport.

   Continue to the next step when your import occurs without errors.

2. The format **ElemNumTextPage** does not yet exist in **testdoc.fm**, so a **FrameMaker Catalog Manager Report** dialog appears, indicating that this format has been created for you. This is expected, and you will refine this format to meet your needs later in this exercise.

Element Catalog Manager Report

April 16, 2026 12:57 pm

Imported EDD: \\vmware-host\Shared Folders\Documents\Books\Fm v18-EDD WB\Workbook Files Master\EDDClassFiles\EDD-WB-Files-2026-18-0-0a\EDD.fm

Destination Document: \\vmware-host\Shared Folders\Documents\Books\Fm v18-EDD WB\Workbook Files Master\EDDClassFiles\EDD-WB-Files-2026-18-0-0a\testdoc.fm

Messages...

Creating new cross-reference format (ElemNumTextPage).

Element Catalog Manager completed.

3. Close the log file.
4. In the **Document View** of **testdoc.fm**, place your cursor in a **Para** element.
5. From the **Elements** panel, insert an **XRef** element.

   If not already visible, the **Cross-Reference** panel appears with **ElemNumTextPage** preselected from the **Format** popup menu.

   The format was created for you on import, but the format's definition is empty (blank right now, but normally displayed below the format name, in the area circled in the screen capture).

   Next, you will edit the definition.

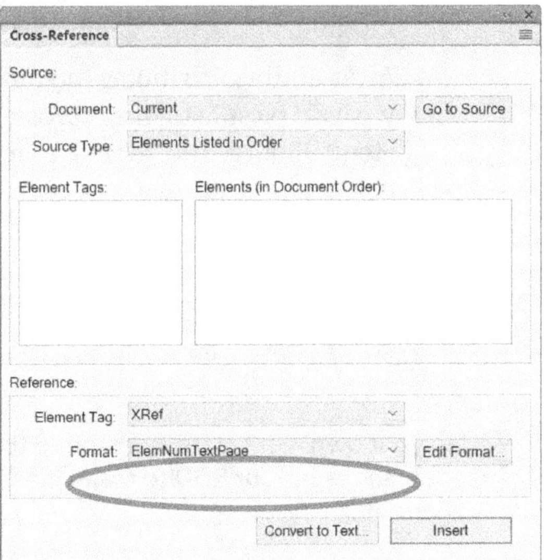

6. Click **Edit Format**.

   The **Edit Cross-Reference Format** dialog appears.

7. In the **Definition** text box, type:
   See <$elemparanum> <$elemtext>,
   on page <$elempagenum>.

8. Click the **Change** button to change the format and store its definition in the cross-reference catalog for this document.

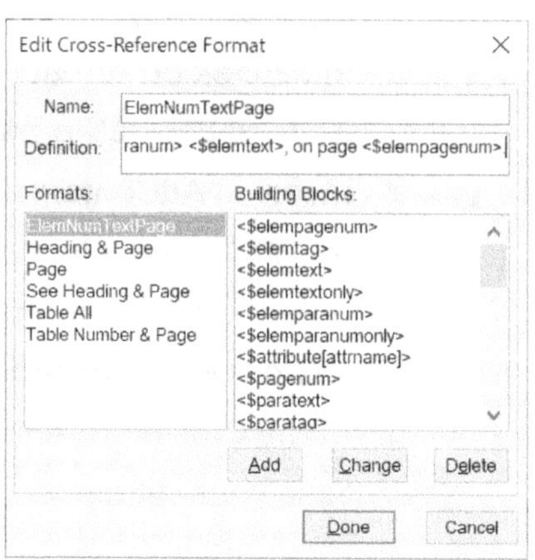

Note: Adding a leading space at the beginning of a cross-reference format allows you to omit a leading space when inserting full-sentence cross-references.

9. Click **Done** to dismiss the **Edit Cross-Reference Format** dialog.

10. Dismiss the **Update References** dialog by clicking **Update** if it appears.

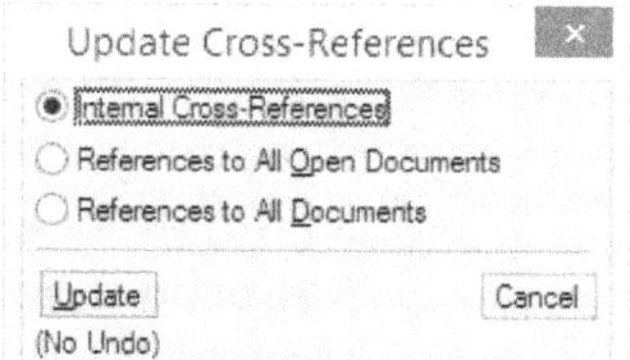

11. Review the **Cross-Reference** panel and note that it is empty.

    Before you can successfully insert an element-based cross-reference, you need to define some attributes for the **XRef** element and the elements you wish to reference. You will do that in a later module.

    After adding attributes to your content model, you will be able to insert cross-references to **Title**, **Head**, **Figure**, and **Table** elements.

12. Save your changes.

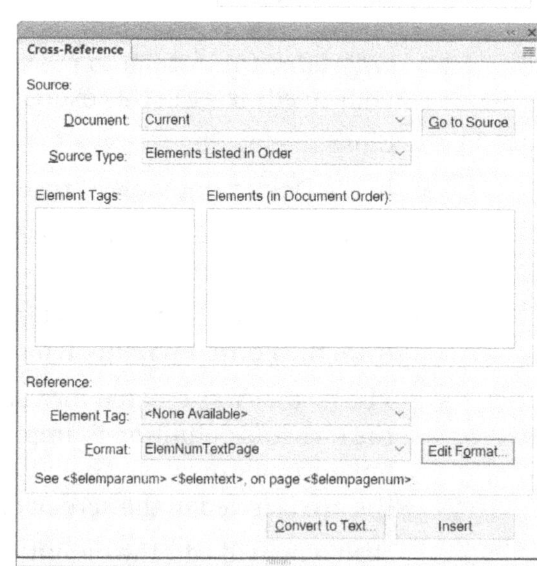

In your own work, consider storing the format of ElemNumTextPage and other unstructured catalog definitions in your EDD to make it easier to import the format into your chapters using File > Import > Formats.

Equation Elements

Exercise 3: Defining an Equation and Specifying Its InitialObjectFormat

In this exercise, you will:

- Redefine the **Section** element to contain optional **EQ** elements
- Define the **EQ** element as an **Equation**, using the **InitialObjectFormat** of **LargeEquation**

LargeEquation indicates the size of the characters in the equation, not the number of characters the equation can have. The user can specify a different equation size without incurring a format rule override.

1. In the EDD, locate the **Section** element definition.
2. In the **Document Window**, add **EQ** to the **GeneralRule** as follows:

   ```
   Head, (Para, (WarnNote | List | Table | EQ)?)+, (Section, Section+)?
   ```

3. Create a new **Element** and tag it **EQ**.
   a. In the **Structure View**, click below the last **Element** definition.
   b. From the **Elements** panel, insert an **Element** element.
      An **Element** and a child **Tag** element appear.
   c. In the **Tag** element, type: `EQ`
4. Define **EQ** as an **Equation**.
   a. Click below the **Tag** element.
   b. From the **Elements** panel, insert **Equation**.
      Equation element appears.
5. From the **Elements** panel, insert **InitialObjectFormat**.
   An **InitialObjectFormat** element appears.
6. Insert **AllContextsRule**.
   An **AllContextsRule** element appears.
7. Insert **LargeEquation**.
   A **LargeEquation** element appears.
8. Save your changes.

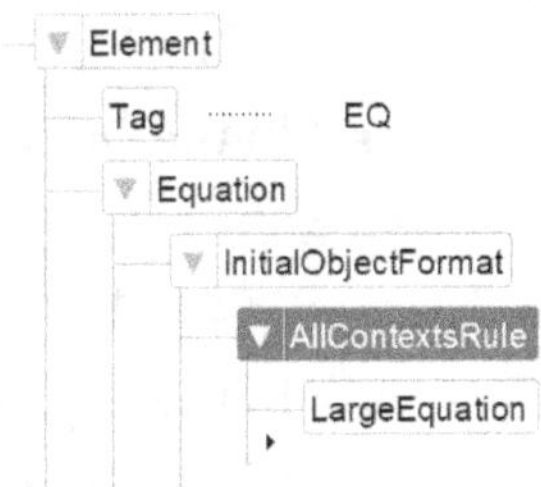

Exercise 4: Reimporting and Testing the Equation Element

In this exercise, you will reimport the EDD into the structured template and test your element definitions for Section and EQ.

1. Reimport the element definitions from your EDD into your test document. If you get errors when importing the element definitions, correct your EDD and reimport.
   Continue to the next step when your import occurs without errors.
2. In the **Structure View**, click below a **Para** element in a **Section**.
3. From the **Elements** panel, insert an **EQ** element.

An equation frame appears with the question mark (?) in the middle of the frame selected.

4. Click/Double-click on the question mark as needed to highlight it and type the following: `x+x=2x`

[[Chapter Title]
[[Section Title]
[Para Text]

$$x + x = 2x$$

5. Save your changes.

Graphic Elements

Exercise 5: Defining a Graphic and Specifying Its InitialObjectFormat

In this exercise, you will:

- Redefine **Section** to contain an optional **Figure** elements
- Redefine **Item** to contain an optional **Figure** elements
- Define **Figure** as a container of **Caption** followed by **Graphic**, with an **AutoInsertion** of **Caption**
- Define **Caption** as a **Container** of <TEXT>
- Define **Graphic** as a **Graphic** with an **InitialObjectFormat** of ImportedGraphic

ImportedGraphic indicates that the **Import File** dialog, rather than the **Anchored Frame** dialog, will open when the **Graphic** element is inserted. The user can choose to close the **Import File** dialog and use the **Anchored Frame** dialog without incurring a format rule override.

1. In the EDD, locate the **Section** element definition.
2. In the **Document Window**, add **Figure** to the **GeneralRule** as follows:

```
Head, (Para, (WarnNote | List | Table | EQ | Figure)?)+,
(Section, Section+)?
```

3. In the EDD, locate the **Item** element definition.
4. In the **Document Window**, rewrite and expand the **GeneralRule** for **Item** as follows:

```
(Para, (WarnNote | Figure)?)+
```

5. Create a new **Element** and tag it **Figure.**
   a. In the **Structure View**, click below the last **Element** element.
   b. From the **Elements** panel, insert an **Element** element.
      An **Element** and a child **Tag** element appear.
   c. In the **Tag** element, type: `Figure`

6. Define **Figure** as a **Container** of **Caption, Graphic.**
   a. Click below the **Tag** element.
   b. From the **Elements** panel, insert a **Container** element.

      A **Container** element and a **GeneralRule** child element appear.
   c. In the **GeneralRule** element, type: `Caption, Graphic`
7. Define **Figure** as having an **AutoInsertion** of **Caption.**

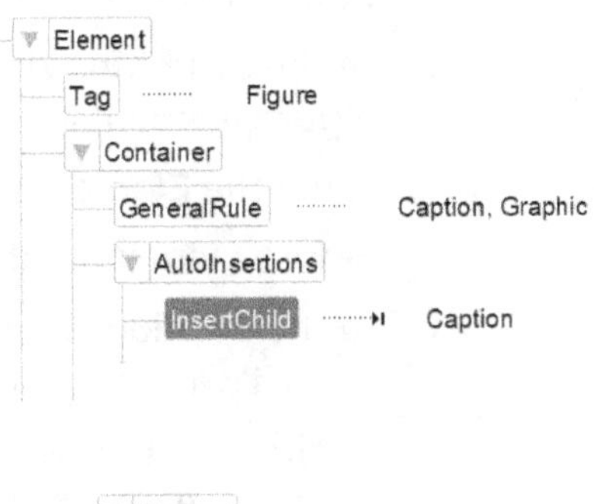

   a. Click below the **Figure's GeneralRule** element.
   b. From the **Elements** panel, insert an **AutoInsertions** element.

      An **AutoInsertions** element with an **InsertChild** element appear.
   c. In the **InsertChild** element, type: `Caption`
8. Create a new **Element** and tag it **Caption.**

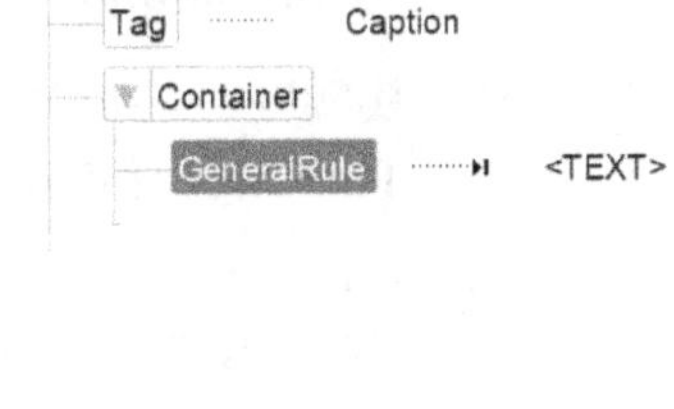

   a. In the **Structure View,** click below the last **Element** in the EDD.
   b. From the **Elements** panel, insert an **Element** element.

      An **Element** and a child **Tag** element appear.
   c. In the **Tag** element, type: `Caption`
9. Define **Caption** as a **Container** of <TEXT>
   a. Click below the **Tag** element.
   b. From the **Elements** panel, insert **Container.**

      Container element and **GeneralRule** child element appear.
   c. In the **GeneralRule** element, type: `<TEXT>`
10. Create a new **Element** and tag it **Graphic.**

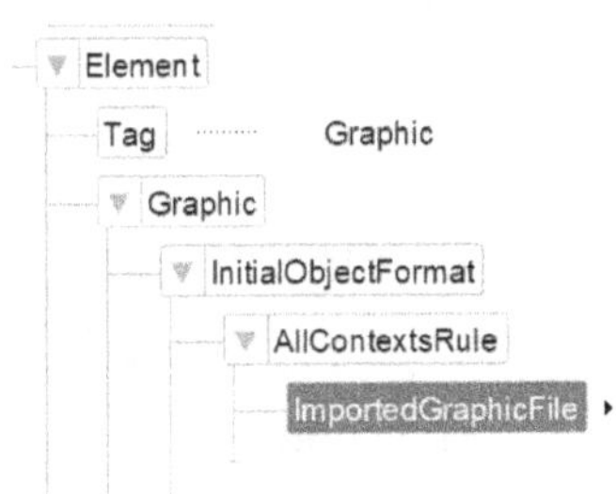

    a. In the **Structure View,** click below the last **Element** in the EDD.
    b. From the **Elements** panel, insert an **Element** element.

       An **Element** and a child **Tag** element appear.
    c. In the **Tag** element, type: `Graphic`
11. Define **Graphic** as a **Graphic.**
    a. Click below the **Tag** element.
    b. From the **Elements** panel, insert a **Graphic** element.
12. From the **Elements** panel, insert **InitialObjectFormat.**

    An **InitialObjectFormat** element appears.
13. Insert an **AllContextsRule** element.
14. Insert an **ImportedGraphicFile** element.
15. Save your changes.

Exercise 6: Reimporting and Testing the Graphic Element

In this exercise, you will reimport the EDD into the structured template and test your element definitions for **Section**, **Item**, **Figure**, **Caption**, and **Graphic**.

1. Reimport the element definitions from your EDD into your test document. If you get errors when importing the element definitions, correct your EDD and reimport.

   Continue to the next step when your import occurs without errors.

2. In the **Structure View**, create a new **Section** to test your **Figure** elements.
3. From the **Elements** panel, insert a **Figure** element.

   Figure element and **Caption** child element appear.

4. In the **Caption** element, type:
   `Caption for Figure`

Section
Head Section for Figure Testing
Para Required Para preceding Figure
Figure
Caption Caption for Figure

5. Click below the **Caption** element, on the line descending from the **Figure** element.
6. Insert a **Graphic** element.

   The **Import** dialog appears.

7. If necessary, change to your class files directory.
8. Select **sidedoor.tif**.
9. If necessary, select **Import by Reference**.

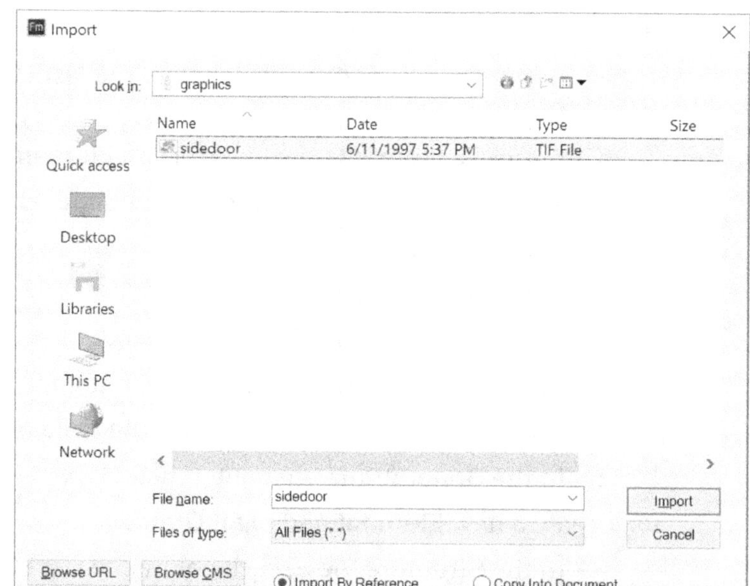

10. Click **Import**.

    The **Imported Graphic Scaling** dialog appears.

11. Select **150 dpi** and click **Set**.

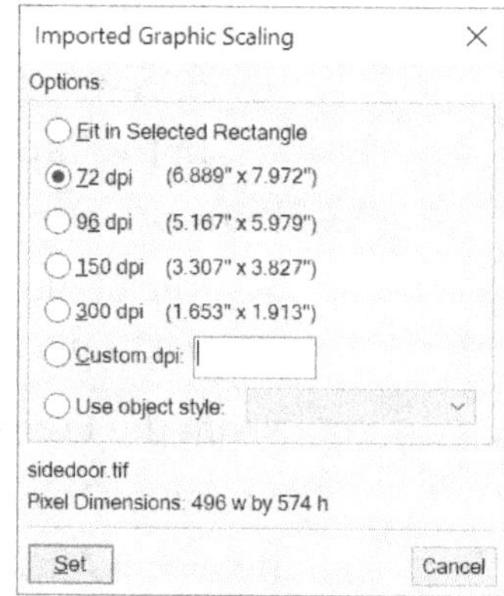

The sample graphic appears below the caption, in an anchored frame sized to fit the graphic.

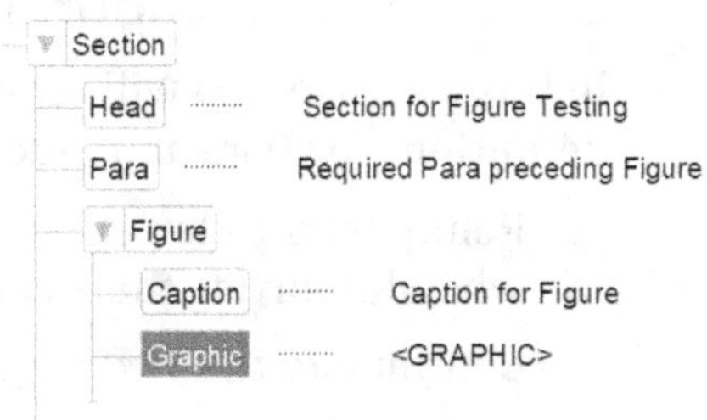

12. Click in any **Item** element but below a **Para** element.
13. Insert a **Figure** element and test by adding caption and **sidedoor.tif** as above.

    A **Figure** element with child elements appears.
14. Save your changes.

Marker Elements

Exercise 7: Defining a Marker and Specifying Its InitialObjectFormat

In this exercise, you will:

- Redefine the **Para** element to contain optional **IndexEntry** elements
- Define the **IndexEntry** element as a **Marker**, using the **InitialObjectFormat** of **Index**

Index will be the preselected marker type in the **Insert Marker** dialog when the user inserts the **IndexEntry** elements. The user can select a different marker type without incurring a format rule override.

1. In the EDD, locate the **Para** element definition.
2. In the **Document Window**, change the **GeneralRule** as follows:

   `(<TEXT> | Footnote | XRef` **| IndexEntry**`)+`
3. Create a new **Element** and tag it **IndexEntry**.
   a. In the **Structure View**, click below the last **Element** element.
   b. From the **Elements** panel, insert an **Element** element.

      An **Element** and a child **Tag** element appear.
   c. In the **Tag** element, type: `IndexEntry`
4. Define **IndexEntry** as a **Marker**.
   a. Click below the **Tag** element.
   b. From the **Elements** panel, insert **Marker**.

5. From the **Elements** panel, insert **InitialObjectFormat**.
6. Insert **AllContextsRule**.
7. Insert **MarkerType**.
8. Insert **Index**.
9. Save your changes.

Exercise 8: Reimporting and Testing the Marker Element

In this exercise, you will reimport the EDD into the structured template and test your element definitions for **Para** and **IndexEntry**.

1. Reimport the element definitions from your EDD into your test document. If you get errors when importing the element definitions, correct your EDD and reimport.

   Continue to the next step when your import occurs without errors.
2. In the **Structure View**, click in any **Para** element.
3. From the **Elements** panel, insert an **IndexEntry** element.

   The **Insert Marker** dialog appears with **Index** preselected from the **Marker Type** popup menu.

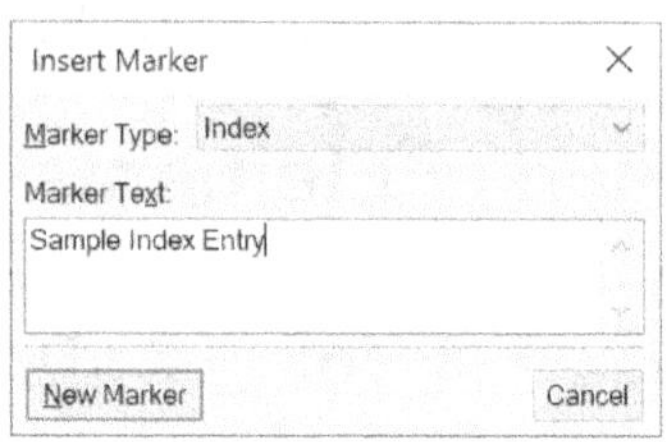

4. In the **Marker Text** dialog, type: `Sample index entry`
5. Click the **New Marker** button to add the marker.

   In the **Structure View**, the **IndexEntry** element appears with "Sample index entry" to the right of the element bubble. Also, if you view your text symbols (**View>Text Symbols**) you will see a marker symbol (**T**) at the insertion point in the **Document Window**.

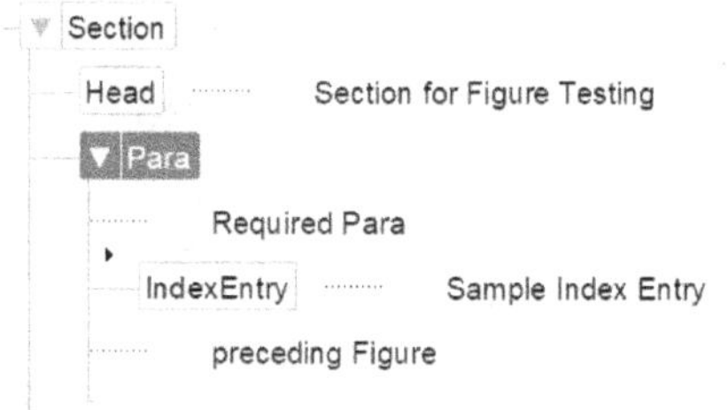

6. Save your changes.

Using a SystemVariableFormatRule

A required part of definitions for SystemVariable elements

- Identifies which system variable to use
- A SystemVariableFormatRule is not a suggestion—users cannot change to another format without creating format override

You can apply a SystemVariableFormatRule with:

- An AllContextsRule—wherever the element appears
- A ContextRule—when element is in certain context
- A LevelRule—when element is nested a specified number of levels in a specified ancestor
- As a DefaultSystemVariable—using the FilenameLong system variable

Exercise 9: Defining a System Variable and Specifying Its Format Rule

In this exercise, you will:

- Expand the **GeneralRule** for the **Para** element to contain a **Date** element
- Define the **Date** element as a **System Variable**, using the **SystemVariableFormatRule** of **CurrentDateLong**.

The **Current Date (Long)** system variable will be inserted when the user inserts the **Date** elements. The user *cannot* select a different system variable without incurring a format rule override.

1. In the EDD, locate the **Para** element definition.

2. In the **Document View**, add **Date** to the **GeneralRule** as follows:

   `(<TEXT> | Footnote | XRef | IndexEntry | Date)+`

3. Create a new **Element** and tag it **Date.**
   a. In the **Structure View**, click below the last **Element** in the EDD.
   b. From the **Elements** panel, insert an **Element** element.

      An **Element** and a child **Tag** element appear.
   c. In the **Tag** element, type: `Date`
4. Define **Date** as a **SystemVariable.**
   a. Click below the **Tag** element.
   b. From the **Elements** panel, insert a **SystemVariable** element.
5. From the **Elements** panel, insert a **SystemVariableFormatRule** element.
6. Insert an **AllContextsRule** element.
7. Insert a **UseSystemVariable** element.
8. Insert a **CurrentDateLong** element.
9. Save your changes.

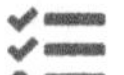

Exercise 10: Reimporting and Testing the System Variable Element

In this exercise, you will reimport the EDD into the structured template and test your element definitions for **Para** and **Date.**

1. Reimport the element definitions from your EDD into your test document. If you get errors when importing the element definitions, correct your EDD and reimport.

   Continue to the next step when your import occurs without errors.
2. In the **Structure View**, click in any **Para** element.
3. From the **Elements** panel, insert a **Date** element.

   Today's date appears at the insertion point.
4. Save your changes.

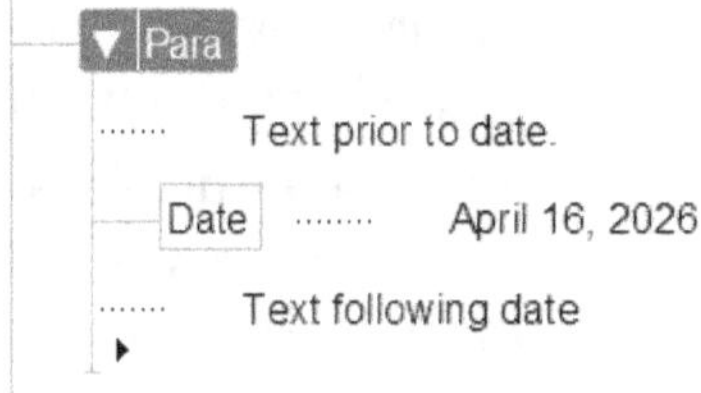

Using Variables on Master Pages

Exercise 11: Inserting a variable on a master page

This exercise doesn't change your EDD. If you don't need practice working with master page formatting, you can choose to skip this exercise and use the start file specified at the start of the next chapter.

Master pages are generally not something that get changed within your content documents. However, it's good for you to know what master pages are, and where a template designer (maybe you...) would go to change them within a template document.

In this exercise, you will open a sample template file and insert a page number variable on the "First" master page.

1. Open the **template.fm** from your class files directory
2. Click on page 1 of your document.
3. Select **View > Master Pages** from the main menu.

   When you view master pages, FrameMaker initially displays the master page associated with the body page containing your insertion point.

   In this document, the second master page, labeled **CoverPage**, is assigned to body page 1, so the master page's name displays in the status area at the bottom of the document window.

4. Double-click on the footer text frame on the master page labeled **CoverPage** to place your insertion point in the footer.
5. Select **View > Pods > Variables** from the main menu.

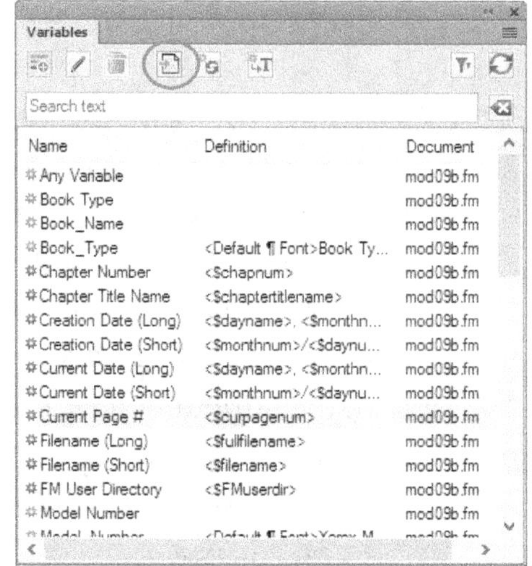

   The **Variables** pod displays.
6. In the **Name** column, select **Current Page #**.

   The variable's definition displays to the right of the Variable scroll list, in the **Definition** column.
7. Click the **Insert** button.

   The "#" symbol displays at the insertion point in the footer text frame. The "#" symbol converts to a page number when viewed on the body pages.
8. Select **View>Body Pages** to return to the "content" pages in your document.

   Page 1 now displays the current page number. In a later exercise you will learn how to update formatting by importing templates into content documents.

1

   In an actual template you would repeat this exercise to place page numbers on other master pages in addition to the **CoverPage** master page.
9. Save and close the **template.fm** file.

Chapter 9: Attribute List

Introduction

This chapter focuses on defining attributes for elements in an **AttributeList**.

Objectives

- Review uses of attributes
- Review basic types of attributes and their parts
- Specify the attribute **Name**
- Specify the attribute **Type**
- Indicate whether the attribute is **Optional** or **Required**
- Indicate whether the attribute is **ReadOnly**
- For attribute of numeric type (**Integer, Integers, Real, Reals**), define a **Range** of values
- For attributes of type **Choice**, define a list of **Choices**
- For attributes indicated as **Optional**, specify a **Default**

Overview

Attributes are an optional part of all element definitions

- An **AttributeList** may contain one or more attributes
- Attributes can be used for general descriptions, cross-referencing, formatting, and prefixing

Potential attributes for metadata:

- A **Class** attribute of **Report** element could identify level of security
- A **Version** attribute of **Chapter** element might identify the current revision
- An **Author** attribute of **Section** element could identify the author of a portion of a document

Examples of attributes for cross-referencing elements:

- Use a **UniqueID** type attribute for all elements to be cross-referenced, such as **Figure, Table, Title, Head**
- User can enter meaningful unique identification for element if not set to **ReadOnly**
- If **UniqueID** has a type attribute of **ReadOnly**, FrameMaker will autogenerate a value
- If **ReadOnly**, defining the **UniqueID** type attribute as **Optional** will prevent a "missing attribute value" validation error if element never referenced
- Use an **IDReference** type attribute for cross-reference elements (**XRef, TableXRef, FigXRef**) that refer to other elements
- Define an **IDReference** type attribute as **Required** to force a link to a **UniqueID** attribute value for validity

- Define an **IDReference** type attribute as **ReadOnly** if you don't want user to enter or edit the value
- If a referenced element has no **UniqueID** value, FrameMaker generates one and then copies the generated **UniqueID** for the **IDReference** attribute value

Examples of attributes used for formatting elements:

Name/value attribute pairs in format rule can drive formatting of:

- Initial table format
- Initial object format
- System variable format
- Paragraph formatting
- Text-range formatting

Examples of attributes used for prefixing elements:

Format rules can use attribute values to provide text for prefix or suffix:

- **Prefix** is text range that appears at beginning of element (before element's content)
- **Suffix** is text range that appears at end of element (after content)

Basic types of attributes and their parts

Defaults can only be provided for attribute values defined as **Optional** (not **Required**). Plural attribute types (**Strings**, **Integers**, **Reals**) may have more than one default value

Attribute name restrictions

- Required part of all attribute definitions
- Type descriptive name of attribute
- Up to 255 characters, but better to keep concise
- Case-sensitive
- Cannot contain white space
- Cannot contain any of these special characters:

  ( ) & | , * + ? < > % [ ] = ! ; : { } "

Attribute types

- Are a required part of all attribute definitions
- You must select one of predefined attribute types

Attributes are either optional or required

- This is a required part of all attribute definitions
- You must specify whether or not attribute requires value for each instance of element
- If an attribute requires but does not have value, FrameMaker identifies attribute as invalid

ReadOnly

- Is an optional part of all attribute definitions
- You may choose whether to restrict users from entering and editing attribute values

Range

- Optional part of attribute definitions for numeric types:

  Integer, Integers, Real, Reals
- Specify whether to restrict users to an inclusive range of values

  Attributes window displays range of values when this attribute is selected

Choices

- **Choices** are a required part of attribute definitions for **Choice** attributes
- **Choices** are typed in manually in the **Choices** element
- The **Attributes** window displays choices in **Choices** popup menu when an attribute is selected
- **Choices:**
  - Can have labels up to 255 characters
  - Can have white-space characters
  - Cannot have any of these special characters:

    ( ) & | , * + ? < > % [ ] = ! ; : { } "

Default

- Defaults are an optional part of attribute definitions for:

  String or **Strings**, **Integer** or **Integers**, **Real** or **Reals**, **Choice**

  But only if defined as **Optional** attribute value (not **Required**)
- Inserts a default value if user elects not to enter a value
- Can specify more than one default value for plural types (**Strings**, **Integers**, **Reals**)

Defining attributes for metadata

Exercise 1: Defining a Required String Attribute

In this exercise, you will define a **Required String** attribute named **Author** for the **Chapter** element.

To do this:

1. If it is not already open, from your class files directory, open **EDD.fm**, the EDD you've been modifying throughout the class.

   If you did not finish the previous chapter's modifications to the EDD, please open **Chapter 9-start Attributes.fm** instead, and save it in your class files directory as **EDD.fm**.

For instructions on downloading class files, see "Downloading class files" on page 1.

2. In the EDD, locate the **Chapter** element definition.
3. Click below the **Chapter's GeneralRule** element (or click below the **ValidHighestLevel** element if you inserted it below, rather than above, the **Chapter's GeneralRule**).
4. From the **Elements** panel, insert an **AttributeList** element.

   An **AttributeList** element with an **Attribute** child element is inserted automatically, along with a nested **Name** child element.

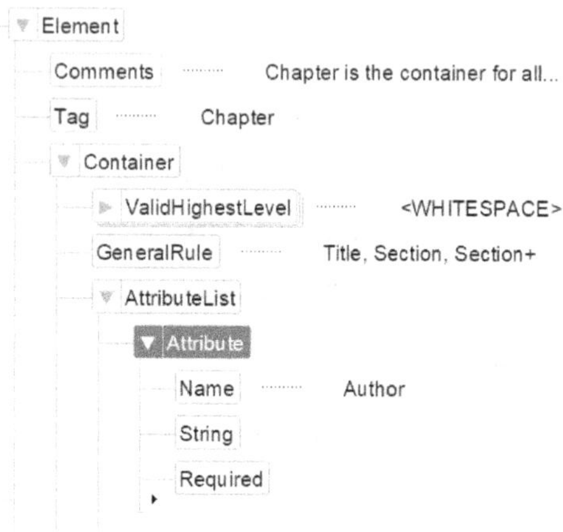

5. In the **Name** element, type: `Author`
6. Click below the **Name** element, on line descending from Attribute element.
7. From the **Elements** panel, insert **String**.

   A **String** element appears with insertion point below it.
8. From the **Elements** panel, insert a **Required** element.
9. Save your changes.

Exercise 2: Defining an Optional Real Attribute with a Range and Default

In this exercise, you will define an **Optional Real** attribute named **Version**, including a **Range** limitation and a **Default** value, for the **Chapter** element.

To do this:

1. In the EDD, locate the **Chapter** element definition.
2. In the **Structure View**, click below the **Author** attribute in the **AttributeList** element.
3. From the **Elements** panel, insert an **Attribute** element.

   The **Attribute** and **Name** child element inserted automatically.

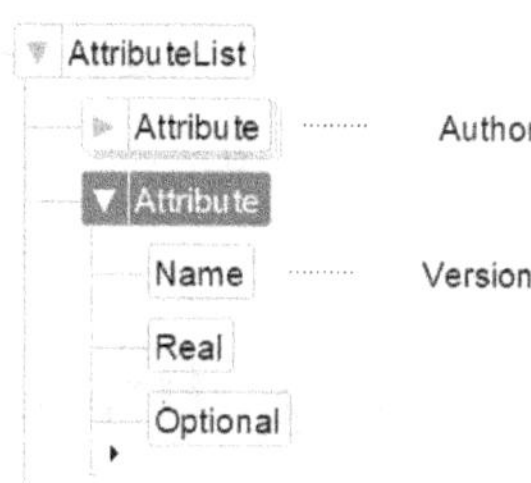

4. In the **Name** element, type: `Version`
5. Click below the **Name** element, on line descending from the **Attribute** element.
6. From the **Elements** panel, insert a **Real** element.
7. From the **Elements** panel, insert an **Optional** element.
8. From the **Elements** panel, insert a **Range** element.

   A **Range** element and **From** child element appear.
9. In the **From** element, type: `1`
10. Click below the **From** element, on line descending from **Range** element.
11. From the **Elements** panel, insert a **To** element.
12. In the **To** element, type: `5`
13. Click below the **Range** element, on the line descending from the **Attribute** element.
14. From the **Elements** panel, insert a **Default** element.

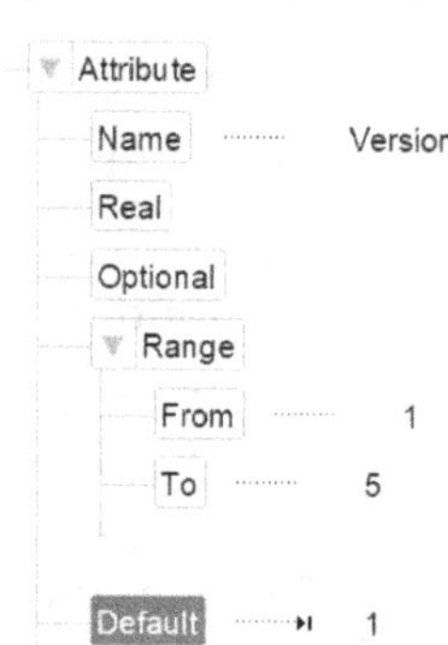

15. In the **Default** element, type: `1`
16. Save your changes.

Exercise 3: Reimporting and Testing Attributes

In this exercise, you will reimport the EDD into the structured template, verify that **Always Prompt for Attribute Values** is turned on, and test your attributes on **Chapter**.

1. Reimport the element definitions from your EDD into your test document. If you get errors when importing the element definitions, correct your EDD and reimport.

   Continue to the next step when your import occurs without errors.

2. Verify that **Always Prompt for Attribute Values** is turned on.

   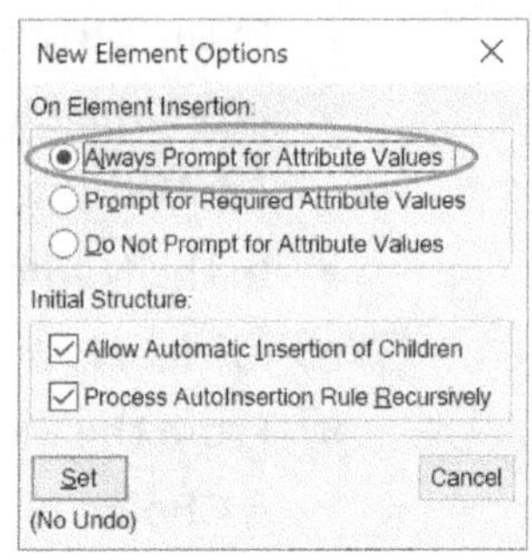

   The author can choose to be prompted for attribute values upon insertion of elements.

   a. From the **Element** menu, choose **New Element Options.**

      The **New Element Options** dialog appears.

   b. If not already on, turn on **Always Prompt for Attribute Values.**

   c. Click **Set.**

3. In the **Structure View**, select and delete the **Chapter** element.

   Since **Chapter** is the highest-level element, all the contents disappear.

4. From the **Elements** panel, insert a **Chapter** element.

   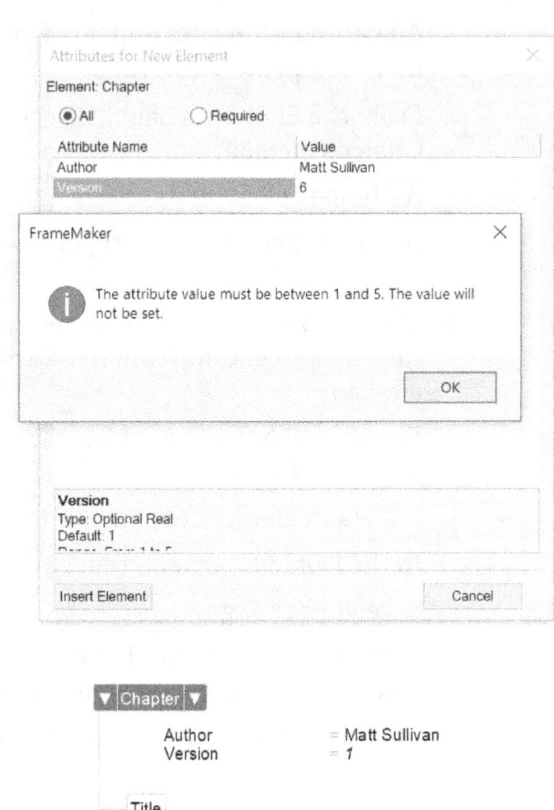

   The **Attributes for New Element** dialog appears.

5. In the **Attribute Name** list, select **Author.**
6. Type a string value for the **Author** attribute and click **Set Value.**
7. In the **Attribute Name** scroll list, select **Version.**

   The dialog displays the properties of the **Version** attribute.

8. Type a number that is not between `1.0` and `5.0` (like 6) and click **Set Value.**

   An alert appears, indicating your value is invalid.

9. Click **OK** to dismiss the alert.
10. Click **Insert Element.**

    The **Version** uses the default value of **1.0.**

11. Save your changes.

Defining Attributes for Prefixes and Formatting

Exercise 4: Defining an optional choice attribute to provide a prefix

In this exercise, you will define an optional choice attribute named **MessageType** with a **Default** value for the **WarnNote** element.

In a later exercise, you will use the value of the **MessageType** attribute to control the prefix of the **WarnNote** element.

To do this:

1. In the EDD, locate the **WarnNote** element definition.
2. In the **Structure View**, click below **WarnNote**'s **GeneralRule** element.
3. From the **Elements** panel, insert **AttributeList.**

   An **Attribute** child element is inserted automatically with a nested **Name** child element.
4. In the **Name** element, type: `MessageType`
5. Click below the **Name** element, on the line descending from **Attribute** element.
6. From the **Elements** panel, insert a **Choice** element.

   A **Choice** element appears with insertion point below it.
7. From the **Elements** panel, insert an **Optional** element.

   An **Optional** element appears with insertion point below it.
8. From the **Elements** panel, insert a **Choices** element.

   A **Choices** element appears.
9. In the **Choices** element, type: `NOTE, WARNING`
10. Click below the **Choices** element, on the line descending from the **Attribute** element.
11. From the **Elements** panel, insert a **Default** element.

    A **Default** element appears.
12. In the **Default** element, type: `NOTE`
13. Save your changes.
14. Reimport the element definitions from your EDD into your test document. If you get errors when importing the element definitions, correct your EDD and reimport.

    Continue to the next step when your import occurs without errors.

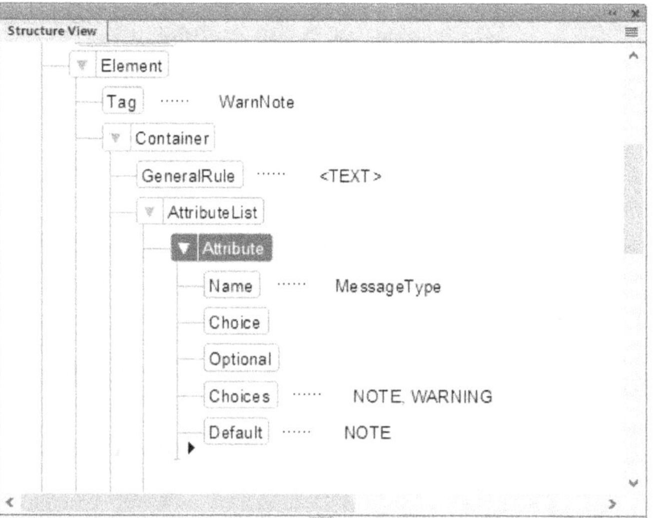

15. From the **Elements** panel, insert **Section, Head,** and **Para** elements to fill out structure requirements.

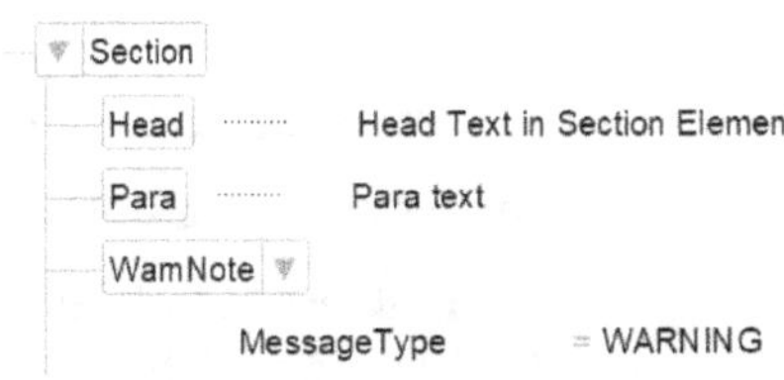

16. From the **Elements** panel, insert a **WarnNote**.

    The **Attributes for New Element** dialog appears.

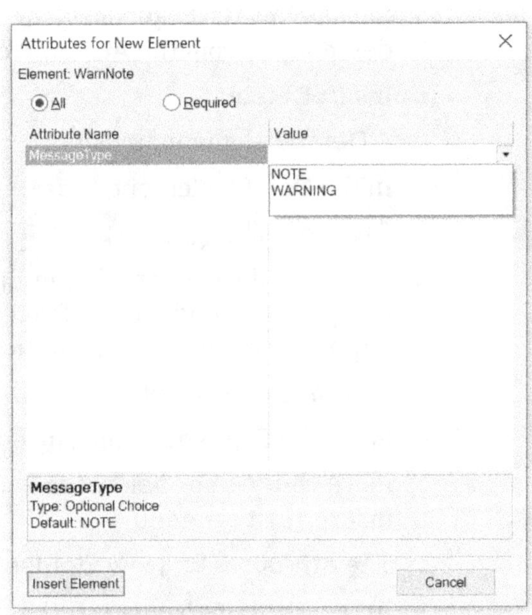

17. In the **Attribute Name** scroll list, select **MessageType.**

    The dialog displays the properties of the **MessageType** attribute.

18. From the attribute value **Choices** popup menu, select **WARNING.**

19. Click **Insert Element.**

    The **Structure View** displays the **MessageType** value. You will use this attribute value later as a prefix for your content.

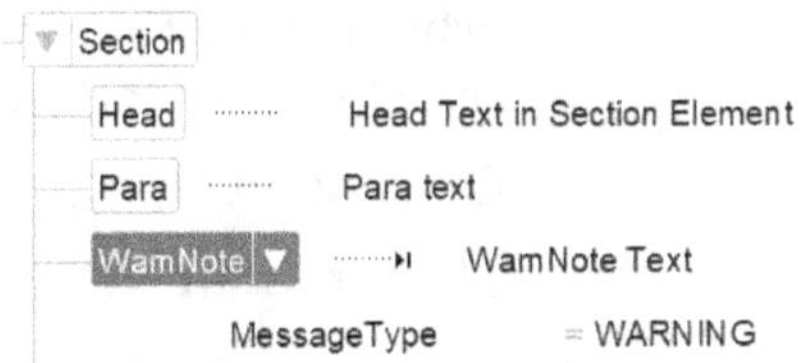

20. Insert text into the **WarnNote** element.

21. Save your changes.

Exercise 5: Defining an Optional Choice Attribute to Provide Formatting

In this exercise, you will define an optional choice attribute named **ListType** for the **List** element and include **Choices** and a **Default** value.

In a later exercise, you will use the value of the **ListType** attribute to control the formatting of the children of the **List** element.

To do this:

1. In the EDD, locate the **List** element definition.
2. In the **Structure View**, click below the **GeneralRule** for the **List** element.
3. From the **Elements** panel, insert an **AttributeList** element.

   An **Attribute** child element is inserted, along with an automatically nested **Name** child element.
4. In the **Name** element, type: `ListType`
5. Click below the **Name** element, on the line descending from the **Attribute** element.

6. Insert a **Choice** element.
7. Insert an **Optional** element.
8. Insert a **Choices** element.
9. In the **Choices** element, type:
   `Bulleted, Numbered`
10. Click below the **Choices** element, on the line descending from the **Attribute** element.
11. Insert **Default**.

    A **Default** element appears.
12. In the **Default** element, type: `Bulleted`
13. Save your changes.
14. Reimport the element definitions from your EDD into your test document. If you get errors when importing the element definitions, correct your EDD and reimport.

    Continue to the next step when your import occurs without errors.

Element
Tag ······· List
Container
GeneralRule ······· Item, Item+
AttributeList
Attribute
Name ······· ListType
Choice
Optional
Choices ······· Bulleted, Numbered
Default ······· Bulleted

15. From the **Elements** panel, insert a **Para** with text at the end of your section.
16. Place your cursor below the **Para** element and insert a **List** element.

    The **Attributes for New Element** dialog appears.

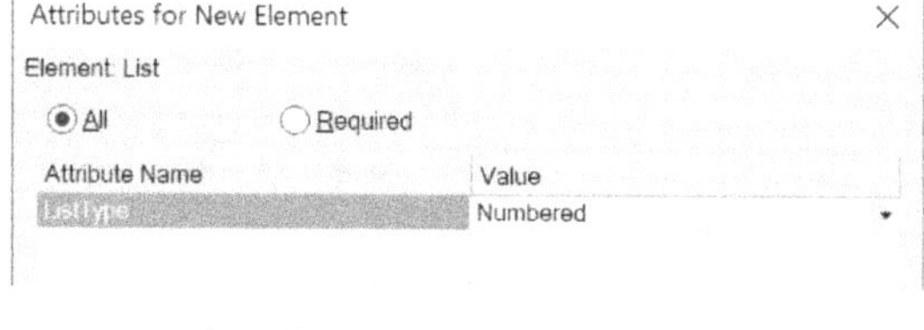

17. In the **Attribute Name** scroll list, select **ListType**.
18. From the attribute value **Choices** popup menu, select **Numbered**. You will set numbering for this list in another lesson.
19. Click **Insert Element**.

    The **Structure View** displays the attribute value.

20. Save your changes.

Exercise 6: Defining an optional choice attribute to provide an InitialObjectFormat

In this exercise, you will define an optional choice attribute named **ReadyToImport** for the **Figure** element and include choices and a default value.

In a later exercise, you will use the value of the **ReadyToImport** attribute to control whether the **Import File** dialog or **Anchored Frame** dialog appear when inserting the **Figure** element's child **Graphic** element.

To do this:

1. In the EDD, locate the **Figure** element definition.
2. In the **Structure View**, click below **Figure's GeneralRule** element.
3. From the **Elements** panel, insert **AttributeList**.

   An **Attribute** child element inserted automatically with nested **Name** child element.

4. In the **Name** element, type: `ReadyToImport`
5. Click below the **Name** element, on line descending from **Attribute** element.
6. Insert **Choice.**
7. Insert **Optional.**
8. Insert **Choices.**
9. In the **Choices** element, type: `Yes, No`
10. Click below **Choices** element, on line descending from **Attribute** element.
11. Insert **Default.**
12. In the **Default** element, type: `No`
13. Save your changes.
14. Reimport the element definitions from your EDD into your test document. If you get errors when importing the element definitions, correct your EDD and reimport.

    Continue to the next step when your import occurs without errors.
15. Add a **Para** element to the end of your **Section.**
16. From the **Elements** panel, insert a **Figure** element.

    The **Attributes for New Element** dialog appears.

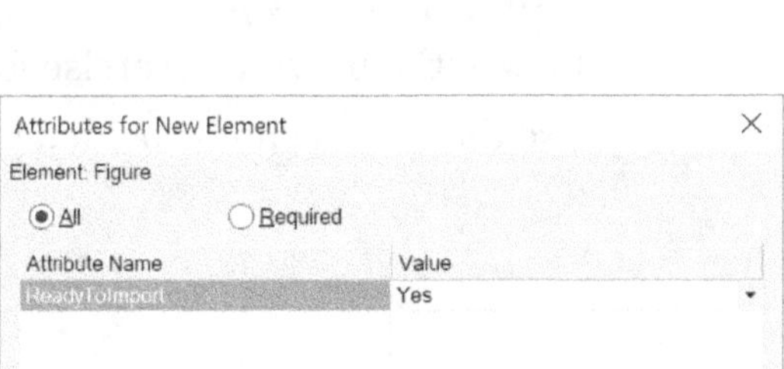
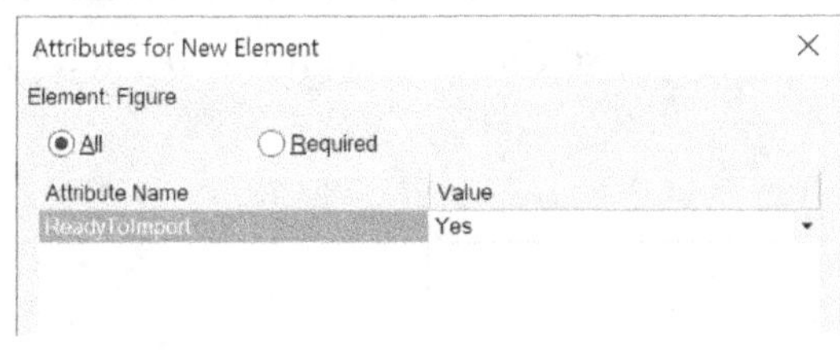

17. In the **Attribute Name** scroll list, select **ReadyToImport** .
18. From the attribute value **Choices** popup menu, select **Yes.**
19. Click **Insert Element.**

    The **Structure View** displays the **ReadyToImport** attribute value.

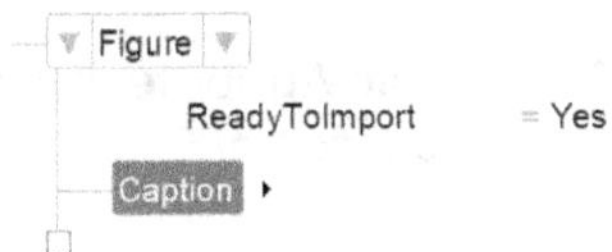

20. Save your changes.

Defining attributes for cross-referencing

Exercise 7: Defining an ID attribute to allow cross-referencing

In this exercise, you will set up attributes for cross-referencing by creating an attribute with a **ReadOnly** specification named **ID** for the **Title, Head, Table,** and **Figure** elements.

In a later exercise, you will use the value of the **ID** for cross-referencing.

To do this:

1. In the EDD, locate the **Title** element definition.
2. In the **Structure View,** click below **GeneralRule** for **Title.**
3. From the **Elements** panel, insert **AttributeList.**

   Nested **AttributeList, Attribute, and Name** elements are inserted automatically.

4. In the **Name** element, type: `ID`
5. Click below the **Name** element, on line descending from **Attribute** element.
6. Insert **UniqueID.**
7. Insert **Optional.**
8. Insert a **Special Attribute Controls** element below **Optional.**
9. Insert a **ReadOnly** element as a child of **SpecialAttributeControls**
10. Save your changes.

Exercise 8: Defining additional ID attributes for cross-referencing

You can use Copy/Paste to speed duplication of common attribute lists and individual attributes.

In this exercise you will add the ID attribute to other elements to enable cross-referencing to them in documents.

To do this:

1. Repeat the previous exercise for the **Head** element.

   Use the screen shot provided for guidance as needed.

While Attribute lists are generally valid following a General Rule element, they must also follow any Inclusion or Exclusion elements.

2. Repeat for the **Table** element.

   Use the screen shot provided for guidance as needed.

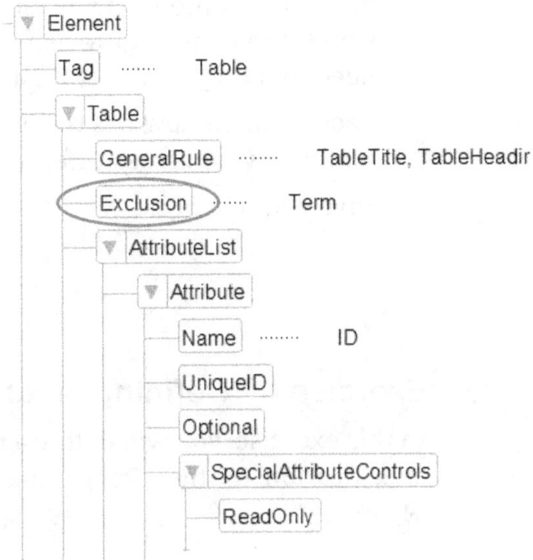

3. Repeat for the **Figure** element.

   Since **Figure** already has an **AttributeList** element, you will add the **ID** attribute as a child to the existing **AttributeList** element.

4. Save your changes.

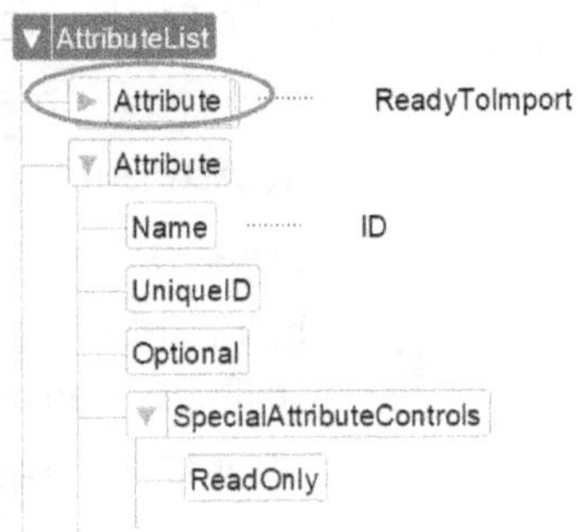

5. Reimport the element definitions from your EDD into your test document. If you get errors when importing the element definitions, correct your EDD and reimport.

   Continue to the next step when your import occurs without errors.

6. From the **Elements** panel, inspect one each of **Title**, **Head**, **Table** and **Figure** elements. Insert elements as needed if they don't yet exist.

   Each element now has the **ID** attribute, which will populate when cross-referenced.
7. Save your changes.

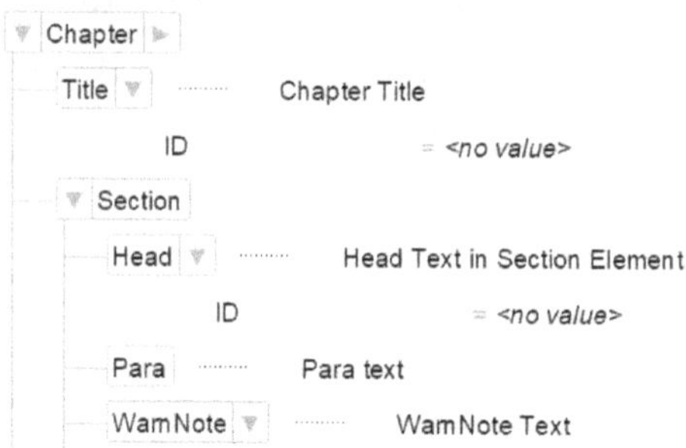

Exercise 9: Defining an attribute to store a cross-reference ID

In this exercise, you will define a required attribute named **IDRef**, including a **ReadOnly** specification, for the **XRef** element.

To do this:

1. In the EDD, locate the **XRef** element definition.
2. In the **Structure View**, click above **XRef's InitialObjectFormat** element.
3. From the **Elements** panel, insert **AttributeList.**

   The **Attribute** child element inserted automatically with a nested **Name** child element.
4. In the **Name** element, type: `IDRef`
5. Click below the **Name** element, on line descending from the **Attribute** element.
6. Insert **IDReference.**

   IDReference element appears with insertion point below it.
7. Insert **Required.**

   A **Required** element appears with insertion point below it.
8. Insert a **Special Attribute Controls** element below **Required.**
9. Insert **ReadOnly.**

   A **ReadOnly** element appears.
10. Save your changes.
11. Reimport the element definitions from your EDD into your test document. If you get errors when importing the element definitions, correct your EDD and reimport.

    Continue to the next step when your import occurs without errors.
12. Position your cursor in a **Para** element.

13. From the **Elements** panel, insert an **XRef** element.

    The **Cross-Reference** dialog appears.

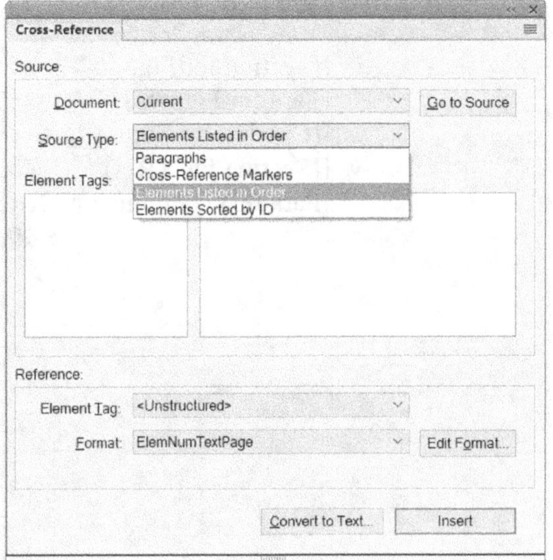

14. From the **Source Type** popup menu, choose **Elements Listed in Order.**

    The **Source Type** scroll list displays all element tags with an attribute of type UniqueID:

    - **Figure**
    - **Head**
    - **Table**
    - **Title**

15. From the **Source Type** scroll list, select **Head.**

    The **Source Text** scroll list displays all occurrences of the **Head** element in the document.

16. In the **Source Text** scroll list, select any one of your **Head** elements.

    ElemNumTextPage is already selected in the **Format** popup menu, because you defined it as the **InitialObjectFormat** for **XRef.**

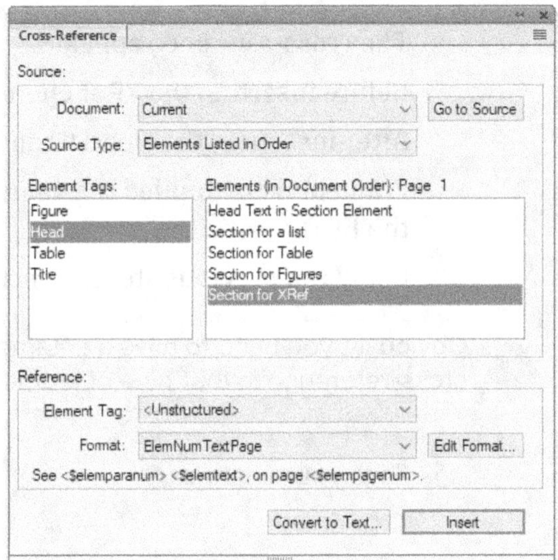

17. Click **Insert**.

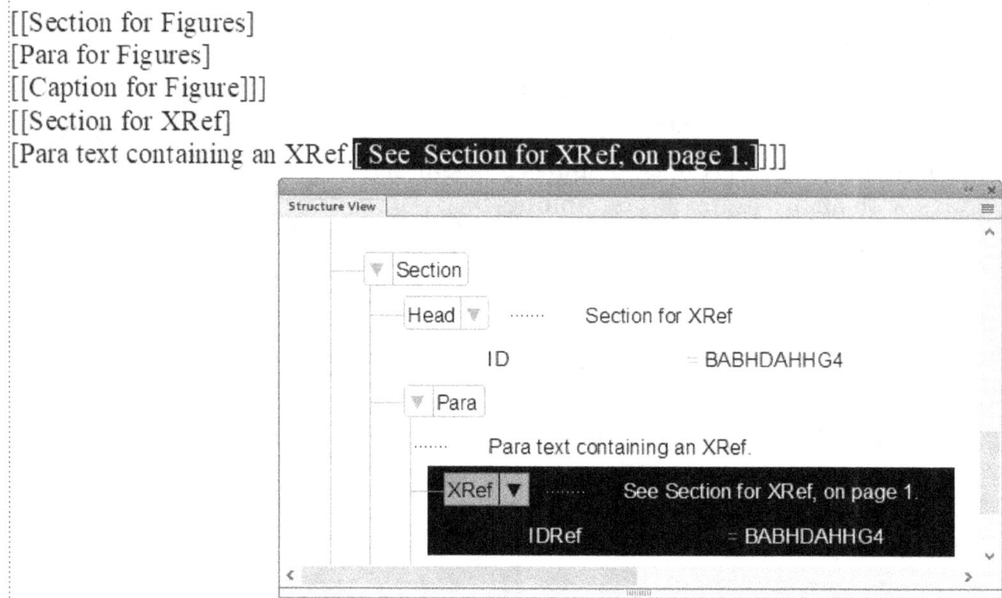

The cross-reference appears.

Before inserting the **XRef** element, the **Section Head** had no **ID**.

After insertion, the **Head ID** and corresponding **XRef IDRef** attributes have matching values.

FrameMaker supplied the value for the **UniqueID** attribute because it was set to **ReadOnly** in the EDD.

18. Using the previous steps, insert **XRef** elements to **Title**, **Table** and **Figure** elements.

Obviously, you need to have an element of a given type in your document before you can insert a cross-reference to that type of element.

19. Save your changes.

Chapter 10: Formatting with AllContextsRules

Introduction

This chapter focuses on writing the **TextFormatRules** for **Container**, **Footnote**, **Table** and table part elements.

Objectives

- Define parts of text format rules
- Specify an **ElementPgfFormatTag**
- Write a(n):
  - **AllContextsRule**
  - **ContextRule** with **If**, **ElseIf**, and **Else** clauses
  - **LevelRule** with **If**, **ElseIf**, **Else** clauses
- Refer to:
  - **ParagraphFormatTag** and **CharacterFormatTag**
  - Individual paragraph/text-range properties
  - **FormatChangeListTag**
- Analyze syntax for naming ancestors, siblings and attribute values
- Define subrules and multiple rules
- Write a **ContextLabel**
- Define a **FormatChangeList/FormatChangeListLimits**

Overview

AllContexts rules are an optional part of element definitions for:

- Container
- Table, TableTitle, TableHeading, TableBody, TableFooting, TableRow, TableCell
- Footnote

AllContexts rules provide formatting for text in elements:

- Rules for **Table**, **TableHeading**, **TableBody**, **TableFooting**, **TableRow** specify formatting only for text in descendant **TableTitle** and **TableCell** elements
- Changes are considered overrides
- When a user imports element definitions into a document, they can keep or remove overrides

The parts that make up TextFormatRules

Part	Description
ElementPgfFormatTag	• Optional, zero or one, before other parts • "Named" base paragraph format stored in document • Defines initial values for all aspects of text and paragraph formatting—basic, default font, pagination, numbering, advanced, table cell
AllContextsRule	• Optional, zero or more, in any order • Formatting changes wherever the element appears
ContextRule	• Optional, zero or more, in any order • With separate clauses: If—one ElseIf—zero or more Else—zero or one • Formatting changes when an element exists in a certain context
LevelRule	• Optional, zero or more, in any order • With separate clauses: If—one ElseIf—zero or more Else—zero or one • Formatting changes when an element is nested within a specified number of levels in an ancestor

Each AllContextsRule, and If/ElseIf/Else clause in a ContextRule or LevelRule can refer to:

- A named **ParagraphFormatTag** or **CharacterFormatTag**
- Individual formatting properties
- A **FormatChangeList**

Text formatting is hierarchical:

- An element can inherit properties from ancestors
- An element can pass on properties to descendants

Order of operations

First, FrameMaker determines which base paragraph format to apply:

- If an element's definition specifies a base paragraph format, then that format is used
- If no base paragraph is specified, FrameMaker searches up through ancestors until it finds element with a base paragraph defined and uses that format
- If FrameMaker reaches the top ancestor of a document without finding a format, FrameMaker uses the default **Body** paragraph format

Defining a base paragraph format for your highest-level element(s) can help avoid confusion later when modifying the EDD.

Then, FrameMaker determines the formatting changes to apply:

- FrameMaker goes back down through hierarchy to the current element, cumulatively picking up formatting changes
- Changes can specify either:
  - Absolute values (fixed value, such as indent expressed as distance from left margin)
  - Relative values (change to current setting, such as amount to move indent)

If the current element is in a table:

- Formatting will not search beyond an ancestor **Table** element
- If no ancestors prior to **Table** element specify paragraph format, the paragraph format stored in the table format will be used

For text in a footnote element:

- No formatting cascade beyond the ancestor **Footnote** element
- If no paragraph formatting is specified within **Footnote** element, main flow footnotes will use the document's current footnote paragraph format and table footnotes will use the current table footnote paragraph format

If the document is part of a book:

- If FrameMaker does not find a paragraph format specification in the content model of a document (when searching outside **Table** or **Footnote**) FrameMaker will continue looking for a paragraph format in ancestor elements of book files
- If found in book hierarchy, FrameMaker uses that paragraph format
- If FrameMaker reaches top of book without finding a paragraph format, FrameMaker uses the default **Body** paragraph format stored in document
- When using the paragraph format from hierarchy in book, FrameMaker goes back down through hierarchy to current element, cumulatively picking up formatting changes

Specifying the ElementPgfFormatTag

Referencing a "named" base paragraph format stored in document

- Defines all properties of text and paragraph formatting—basic, default font, pagination, numbering, advanced, table cell
- If an instance of an element contains text, the text's formatting will initially be the paragraph format settings plus any changes specified for the element's current context in the element's format rules
- The paragraph format of an ancestor is also passed on to element's descendants—until a descendant provides different format

- Rules for **Table**, **TableHeading**, **TableBody**, **TableFooting**, **TableRow** elements pass along formatting only for text in descendant **TableTitle** and **TableCell** elements

Paragraph tags and export to HTML5

FrameMaker maps paragraph styles directly to HTML elements when publishing for devices. FrameMaker allows formatting to be applied to output and also allows formatting specified in the document to pass through into online output. However, FrameMaker requires paragraph styles be in place in order to map to HTML structural elements like **<h1>** and **<h2>**.

For examples of paragraph style mapping, look at FrameMaker's implementation (and publishing) of DITA, which favors application of paragraph styles via the EDD, rather than directly defining formatting values.

Exercise 1: Specifying a Base Tag for the Entire Flow

In this exercise, you will specify an **ElementPgfFormatTag** of **Body** for the **Chapter** element.

The paragraph tag **Body** will be inherited by all elements in the entire structure, unless another element specifies its own paragraph tag. Any individual format changes (such as a weight of bold or size of 14 points) defined in **Chapter** or its descendants will make changes to the properties defined in and inherited from the **Body** paragraph tag.

To do this:

1. If it is not already open, from your class files directory, open **EDD.fm**, the EDD you'll be modifying throughout the class.

   If you did not finish the previous chapter's modifications to the EDD, please open **Chapter 10-start Formatting.fm** instead, and save it in your class files directory as **EDD.fm**.

For instructions on downloading class files, see "Downloading class files" on page 1.

2. In the EDD, locate the **Chapter** element definition.
3. In the **Structure View**, click below the **AutoInsertions** element in the **Chapter** element.

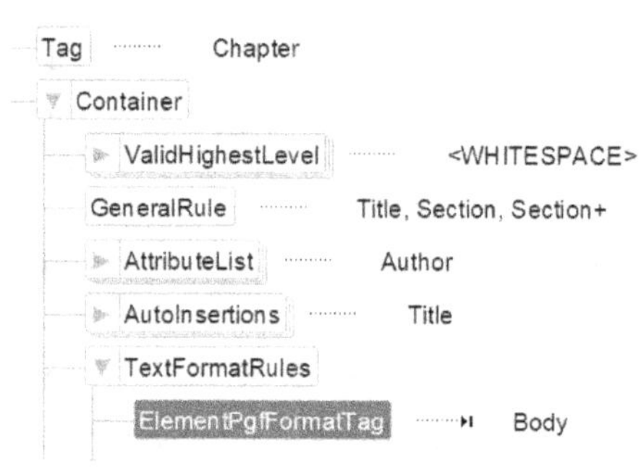

4. From the **Elements** panel, insert a **TextFormatRules** element.

   A **TextFormatRules** element appears.
5. Insert an **ElementPgfFormatTag.**

   An **ElementPgfFormatTag** element appears.
6. In **ElementPgfFormatTag** element, type: `Body`
7. Save your changes.
8. Reimport the element definitions from your EDD into your test document. If you get errors when importing the element definitions, correct your EDD and reimport.

   Continue to the next step when your import occurs without errors.
9. Confirm that all elements now explicitly use the **Body** paragraph format.

Understanding an AllContextsRule

An **AllContextsRule** specifies that formatting will apply to an element in all contexts in which that element can occur. These rules can specify:

- A **ParagraphFormatTag** (or **CharacterFormatTag** if a text range) stored in the document catalogs
- Individual paragraph or text-range formatting properties
- A **FormatChangeList**

When an **AllContextsRule** refers to individual formatting properties, clauses can describe changes to any property in either the **Paragraph Designer** (other than the **Next Paragraph Tag**, which is not used in structured documents) or the **Character Designer** (which is basically the same as **PropertiesFont** properties of the **Paragraph Designer**).

Here is an overview of property organization:

PropertiesBasic	Set indentation, line spacing, paragraph alignment, paragraph spacing, and tab stops
PropertiesFont	Set font, size, and style of text in element
PropertiesPagination	Define placement of paragraph on page and determine how to break paragraph across columns and pages
PropertiesNumbering	Specify syntax and format for automatically generated string, such as number that appears at beginning of procedure step
PropertiesAdvanced	Set hyphenation and word spacing options and determine whether to display graphic with paragraph
PropertiesTableCell	Customize margins of cells and vertical alignment of text in them

Some properties require directly typed values, while other also allow insertion of elements that represent choices (such as **Bold** or **Yes**) from predefined lists.

With some numeric values, you can type either:

- Relative values (positive or negative)—added to current value to set new value
- Absolute values—which override the current value for a property

FormatChangeList

You can use a **FormatChangeList** to:

- Describe set of changes to format properties
- Reduce repetitive entry of the same changes when they are used in multiple locations throughout the EDD

You can refer to a **FormatChangeList** by name in:

- Text format rules—**AllContextsRule**, **ContextRule**, **LevelRule**
- **PrefixRules** or **SuffixRules**
- **FirstParagraphRules** or **LastParagraphRules**

Exercise 2: Specifying an AllContextsRule Referring to "Basic" Properties

In this exercise, you will specify an **AllContextsRule** for the **Chapter** element, that specifies the space above and below the element.

These changes to the **Chapter** element will be inherited by all elements in the entire structure, unless overridden by another element's formatting.

To do this:

1. In the EDD, locate the **Chapter** element definition.
2. In the **Structure View**, click below the **ElementPgfFormatTag** of the **Chapter** element, on the line descending from **TextFormatRules.**
3. From the **Elements** panel, insert an **AllContextsRule** element.
4. To specify paragraph formatting, rather than text-range formatting, insert a **ParagraphFormatting** element.
5. Insert a **PropertiesBasic** element.
6. Insert a **ParagraphSpacing** element.
7. Insert a **SpaceAbove** element and type: `10 pt`
8. Click below the **SpaceAbove** element, on the line descending from the **ParagraphSpacing** element.
9. Insert a **SpaceBelow** element and type: `10 pt`
10. Save your changes.
11. Reimport the element definitions from your EDD into your test document. If you get errors when importing the element definitions, correct your EDD and reimport.

    Continue to the next step when your import occurs without errors.
12. Confirm the addition of 10 points of space above and below each element.

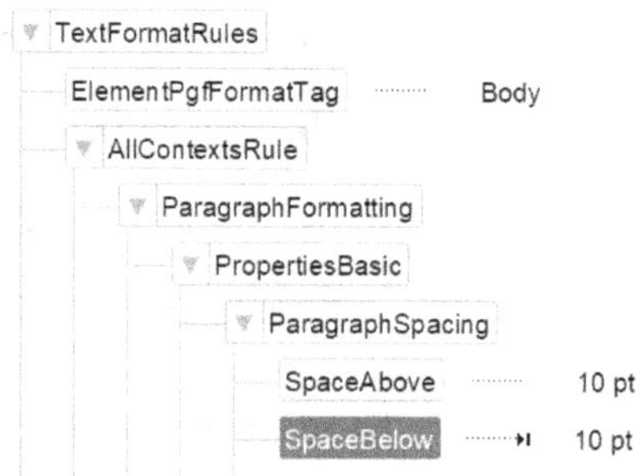

Formatting cascades in structured documents. Because **Chapter**, the highest level ancestor element in this system is the only place where space above and below are defined, all elements have 10 points of space above and 10 points of space below, until you specify different values on child elements.

[[Chapter Title]

[[Head Text in Section Element]

[Para text]

[WarnNote Text]]

[[Section for a list]

[Para describing List]

Exercise 3: Specifying an AllContextsRule Referring to Many Properties

In this exercise, you will specify an **AllContextsRule** for the **Title** element that changes the space below to an absolute value of 50 points, makes the text italic, sets the size to 20 points, and sets angle to italic.

You'll also add an autonumber and add a reference frame called **Double Line** under the **Title** element.

To apply the formatting, do this:

1. In the EDD, locate the **Title** element definition.
2. In the **Structure View**, click below the **AttributeList** of the **Title** element.
3. From the **Elements** panel, insert **TextFormatRules** element.
4. Insert **AllContextsRule** element.
5. Insert **ParagraphFormatting** element.
6. Specify **SpaceBelow** of `50 pt.`
   a. Insert a **PropertiesBasic** element.
   b. Insert a **ParagraphSpacing** element.
   c. Insert a **SpaceBelow** and type: `50 pt`
7. Specify an **Angle** of **Italic** and a **Size** of 20 points.
   a. Click below the **PropertiesBasic** element, on the line descending from the **ParagraphFormatting** element.
   b. Insert a **PropertiesFont** element.
   c. Insert an **Angle** element.

      The **Angle** element and an **Italic** child element appear.
   d. Click below the **Angle** element.
   e. Insert a **Size** element and type: `20 pt`
8. Specify a **Double Line** frame below.
   a. Click below the **PropertiesFont** element, on the line descending from the **ParagraphFormatting** element.
   b. Insert a **PropertiesAdvanced** element.
   c. Insert a **FrameBelow** element and type: `Double Line`

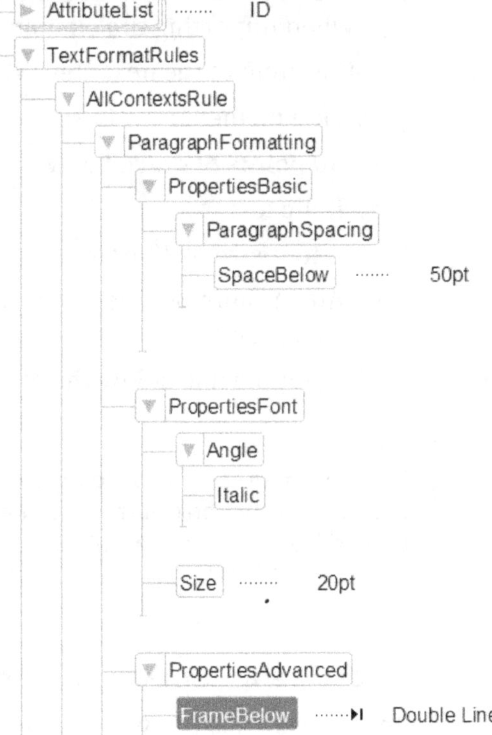

Double Line is the label of a existing frame on the reference page in your structured template. You can specify any frame defined on graphics reference page

9. Specify an **AutonumberFormat** of `C:Chapter <$chapnum>.` followed by a space
   a. Click below the **PropertiesAdvanced** element, on the line descending from the **ParagraphFormatting** element.
   b. Insert **PropertiesNumbering**.
   c. Insert **AutonumberFormat** and type: `C:Chapter <$chapnum>.` followed by a space.

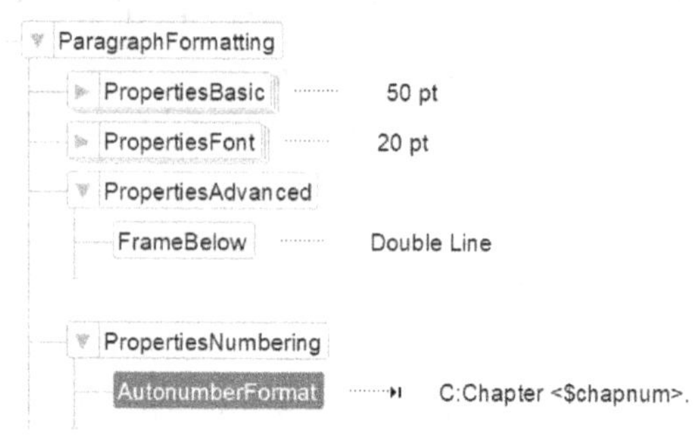

10. Save your changes.
11. Reimport the element definitions from your EDD into your test document. If you get errors when importing the element definitions, correct your EDD and reimport.

    Continue to the next step when your import occurs without errors.

The Title element now has:

- 50 points of space below
- Italic text
- A font size of 20 points
- An autonumber displaying as "Chapter 1. "
- A double line below the element

Chapter 1. [[Chapter Title]

[[Head Text in Section Element]

[Para text to contain an XRef[See Head Text in Section

Many autonumbering building blocks are defined in the Version 18.0 EDD, but <$chapnum> isn't one of them, and must be entered manually. Submit a feature request for this (or anything else you think should be added to FrameMaker) directly to Adobe product managers at https://tracker.adobe.com

Exercise 4: Specifying an AllContextsRule Referring to a FormatChangeList

In this exercise, you will specify an **AllContextsRule** for the **Head** element, that refers to a **FormatChangeList** called **HeadTitleText**

To do this:

1. In the EDD, locate the **Head** element definition.
2. In the **Structure View**, click below the **AttributeList** for the **Head** element.
3. From the **Elements** panel, insert **TextFormatRules**.
4. Insert **AllContextsRule**.
5. Insert **ParagraphFormatting**.
6. Insert **FormatChangeListTag**.
7. In **FormatChangeListTag** element, type: `HeadTitleText`
8. Save your changes.

Exercise 5: Defining a FormatChangeList

In this exercise, you will define a **FormatChangeList** called **HeadTitleText** which specifies a **Weight** of **Bold** and **SpaceAboveChange** of 5 pt.

Elements referencing this list of format changes (like the **Head** element) will be bold, with an additional 5 points of space above them, added to whatever space they have already inherited.

To do this:

1. In the EDD, place your cursor at the end of the **ElementCatalog** element.

While FormatChangeList elements can exist in the same locations as Element definitions, for ease of use, it can be convenient to store **FormatChangeList** elements at the end of the EDD, below all **Element** definitions.

2. From the **Elements** panel, insert **FormatChangeList.**

   A **FormatChangeList** element and a **Tag** child element appear.
3. In the **Tag** element, type: `HeadTitleText`
4. Specify a **Weight** of **Bold.**
   a. Click below the **Tag** element, on line descending from the **FormatChangeList** element.
   b. Insert a **PropertiesFont** element.
   c. Insert a **Weight** element.
   d. Insert a **Bold** element.
5. Specify a **SpaceAboveChange** of `+5 pt`.
   a. Click below the **PropertiesFont** element, on line descending from the **FormatChangeList** element.
   b. Insert a **PropertiesBasic** element.
   c. Insert a **ParagraphSpacing** element.
   d. Insert a **SpaceAboveChange** element and type: `+5 pt`

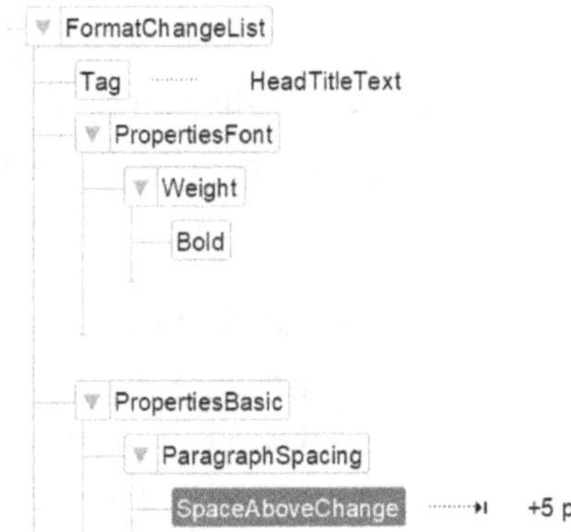

6. Save your changes.
7. Reimport the element definitions from your EDD into your test document. If you get errors when importing the element definitions, correct your EDD and reimport.

   Continue to the next step when your import occurs without errors.

All the **Head** elements are now bold and have an additional 5 points of space above, for a total of 15 pt.

Exercise 6: Specifying a Second AllContextsRule

In this exercise, you will specify a second **AllContextsRule** for your **Title** element. The second rule will refer to the **FormatChangeList**, making the text bold and applying the 5 points of space above the element that you defined in the last exercise.

A **FormatChangeList** applies only the formatting properties applicable to the particular context. Because the **Title** element is the first element containing text in the entire structure, **Title** is at the top of the text column. White space above a paragraph at the top of a text column is ignored. Therefore, the **FormatChangeList** will only apply bold, not any additional space to the **Title** element.

1. In the EDD, locate the **Title** element definition.
2. In the **Structure View**, click below the **AllContextsRule** for the **Title** element, on the line descending from **TextFormatRules**.
3. Refer to the **FormatChangeList** tagged **HeadTitleText**.
   a. Insert an **AllContextsRule** element
   b. Insert a **ParagraphFormatting** element
   c. Insert a **FormatChangeListTag** element
   d. In FormatChangeListTag element, type: `HeadTitleText`

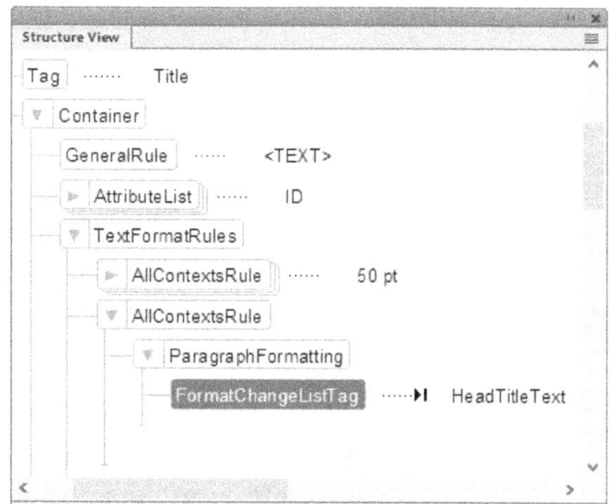

4. Save your changes.
5. Reimport the element definitions from your EDD into your test document. If you get errors when importing the element definitions, correct your EDD and reimport.

   Continue to the next step when your import occurs without errors.

The **Title** element is now bold, but does not have an additional 5 points of space above, because it is positioned at the top of the text column.

Exercise 7: Specifying an AllContextsRule Referring to Text-Range Properties

In this exercise, you will specify an **AllContextsRule** for the **Term** element, identifying it as a text range with a font change of italics.

Defining **Term** in the EDD as a text range will allow it to be embedded within a paragraph-type element without breaking the text into multiple paragraphs.

To do this:

1. In the EDD, locate the **Term** element definition.
2. In the **Structure View**, click below the **GeneralRule** of the **Term** element.
3. From the **Elements** panel, insert a **TextFormatRules** element.
4. Insert an **AllContextsRule** element.

   An **AllContextsRule** element appears.

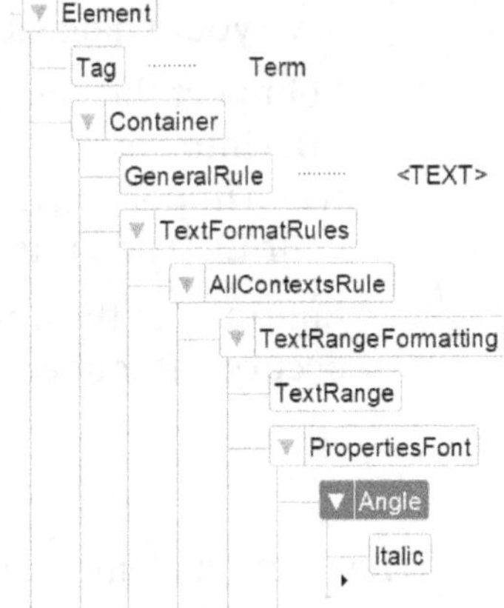

5. To specify text-range formatting, rather than paragraph formatting, insert a **TextRangeFormatting** element.

   A **TextRangeFormatting** element and **TextRange** child element appear.
6. Specify an **Angle** of **Italic**.
   a. Insert a **PropertiesFont** element.
   b. Insert an **Angle** element.

      An **Angle** element and **Italic** child element appear.
7. Save your changes.
8. If needed, add a **Term** element to your test doc.
9. Reimport the element definitions from your EDD into your test document. If you get errors when importing the element definitions, correct your EDD and reimport.

   Continue to the next step when your import occurs without errors.

The **Term** element is now italic, and no longer causes a break when inserted in a paragraph

[[Head Text in Section Element]

[Para text with a [*term*] followed by more text]

Exercise 8: Formatting Captions with Two Rules

In this exercise, you will format the **Caption** for the **Graphic** (both children of **Figure**) by specifying **AllContextsRules** to refer to a **FormatChangeList** and to an autonumbering string.

To do this:

1. In the EDD, locate the **Caption** element definition.
2. In the **Structure View**, click below **GeneralRule** for the **Caption** element.
3. Insert **TextFormatRules.**
4. Attempt the rest on your own, referring to the picture.
5. Save your changes.
6. Reimport the element definitions from your EDD into your test document. If you get errors when importing the element definitions, correct your EDD and reimport.

   Continue to the next step when your import occurs without errors.

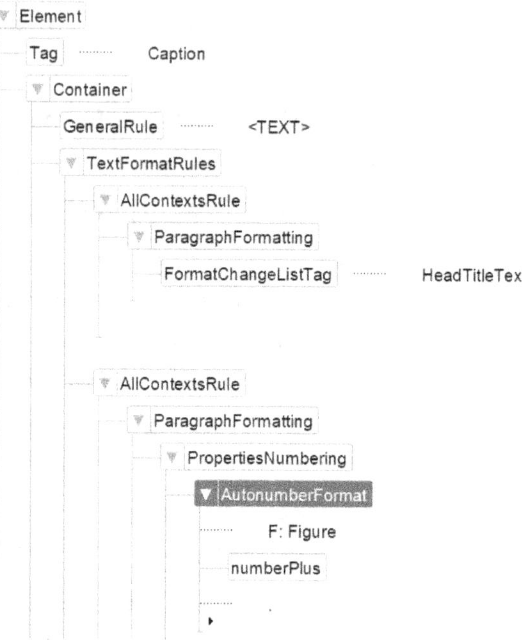

The Caption element is now bold, and has an additional 5 points of space above, for a total of 15 pt.

It also has an autonumber prefix displaying as "Figure 1."

Exercise 9: Formatting TableTitle Text

Tables do not inherit their formatting from their ancestors. Table parts inherit formatting from their ancestor table parts, only up to and including the **Table** element.

For example, formatting specified for:

- A **Table** element passes to all descendants of the **Table** element
- A **TableTitle** element passes to text and children of **TableTitle**
- A **TableHeading** element passes to **TableRow** elements in the **TableHeading** and any **TableCells** in the **TableRows** of **TableHeading** only
- A **TableBody** element passes to **TableRow** elements in the **TableBody** and any **TableCells** in the **TableRows** of **TableBody** only
- A **TableFooting** element passes to **TableRow** elements in the **TableFooting** and any **TableCells** in the **TableRows** of **TableFooting** only
- A **TableRow** element passes to its **TableCell** elements
- A **TableCell** element passes to text and children of **TableCell**

Without format rules, text in the table is formatted according to the properties of the table format specified in its **InitialTableFormat** (or by any other table format the user chooses from the **Insert Table** dialog upon inserting the **Table** element). Table formats have default paragraph formats defined for use in the table title and cells of the table: **TableTitle**, **CellHeading**, **CellBody**, **CellFooting**. In this exercise, you will format the **TableTitle** for the **Table** with one **AllContextsRule** specifying left alignment and autonumbering of `T:Table <n+>. ` (followed by a space).

To do this:

1. In the EDD, locate the `TableTitle` element definition.
2. In the **Structure View**, click below the **GeneralRule** of the **TableTitle** element.
3. Insert **TextFormatRules**.
4. Attempt the rest on your own, referring to the picture.
5. Save your changes.
6. If needed, insert a table into your test document.
7. Reimport the element definitions from your EDD into your test document. If you get errors when importing the element definitions, correct your EDD and reimport.

   Continue to the next step when your import occurs without errors.

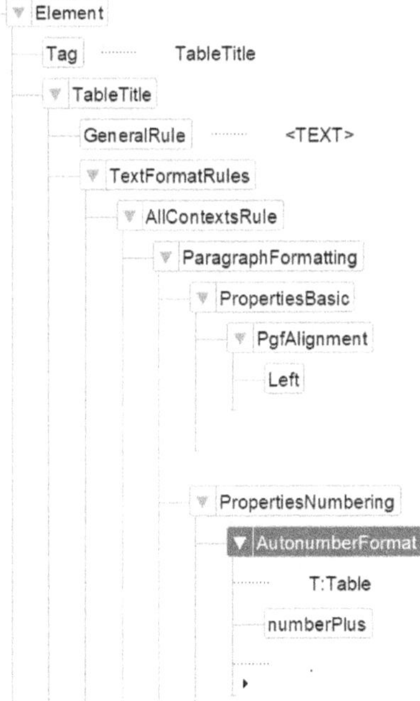

The **TableTitle** element is now left-aligned, with an autonumber displaying as "**Table 1.**"

Table 1. Table Title

Chapter 11: ContextRule formatting rules

Introduction

Objectives

- Write a **ContextRule** with **If**, **ElseIf**, and **Else** clauses
- Write a **LevelRule** with **If**, **ElseIf**, and **Else** clauses
- Analyze syntax for naming ancestors, siblings and attribute values
- Define sub rules and multiple rules
- Write a **ContextLabel**
- Define **FormatChangeListLimits**

Overview

Context rules are an optional part of element definitions for:

- Container
- Table, TableTitle, TableHeading, TableBody, TableFooting, TableRow, TableCell
- Footnote

Context rules identify formatting for text in an element:

- Rules for **Table**, **TableHeading**, **TableBody**, **TableFooting**, **TableRow** specify formatting only for text in descendant **TableTitle** and **TableCell** elements
- Changes are considered overrides
- When user reimports element definitions, user can choose to keep or remove overrides

Writing a ContextRule

Context rules can define one or more possible contexts, with format for each context. They can have separate clauses for different possibilities:

- **If**—only one
- **ElseIf**—zero or more
- **Else**—zero or one

Each **If**, **ElseIf**, and **Else** clause can refer to:

- Named **ParagraphFormatTag** or **CharacterFormatTag** (if text range) stored in document
- Individual paragraph or text-range formatting properties
- **FormatChangeList**

ContextRules that identify ancestors

Often names just a parent

Consider this context rule regarding an Item whose parent is a list:

```
Element (Container): Item
    General rule: <TEXT>
    Text format rules
        1.  If context is: List
                Numbering properties
                    Autonumber format: \b\t
```

Put another way, if an **Item** has a parent of **List**, (independent of the **List**'s ancestors), apply the autonumber format.

Significance of less-than-sign (<) for list of ancestors

Consider this rule, which formats an item based upon the context of the item's parent:

```
Element (Container): Item
    General rule: <TEXT>
    Text format rules
        1.  If context is: List < Preface
                Numbering properties
                    Autonumber format: \b\t
                    Character format: bulletsymbol
            Else, if context is: List < Chapter
                Numbering properties
                    Autonumber format: <n+>\t
```

In this case, if an **Item** is within a **List** within a **Preface** element (regardless of a **Preface**'s ancestors) then bullets are applied to the **Item**

Else if the **Item** is within a **List** within a **Chapter** element (regardless of the **Chapter**'s ancestors) then the Item is formatted as a numbered list.

Similarly, in this next example, something must occur if a **Section** has two ancestors of **Section** (is a 3rd-level section).

If **Section** is within two or more **Section** elements

```
Element (Container): Section
    General rule: Head, Para+
    Text format rules
        1.  If context is: Section < Section
```

Asterisk (*) for unspecified number of successive ancestors

- This example shows the notation for invoking formatting on the 3rd or greater level of **Section**
- If **Section** has parent of **Section** and another ancestor tagged **Section** any number of levels up (grandparent, great-grandparent)

```
Element (Container): Section
    General rule: Head, Para+
    Text format rules
```

1. If context is: Section < * < Section

OR indicators (|) to test specification for any ancestor in group

Occurance indicators can help combine formatting statements. In this example, Items within Lists that are in *either* Preface or Chapter elements are formatted by this rule. Of course, the rule is shown, but the formatting associated with the rule is omitted.

Element (Container): Item
General rule: <TEXT>
Text format rules
1. If context is: List < (Preface | Chapter)

So, if an **Item** is in a **List** that is within either a **Preface** or a **Chapter**, the text formatting specified would occur.

Exercise 1: Specifying a ContextRule with One Clause Naming a Parent

In this exercise, you will specify a **ContextRule** for the **Section** element that indents a **Section** (and all its descendants) when nested within a parent **Section**.

Because you are using the "change" and "relative" elements, each nested level of **Section** will be indented .5 inches further than the previous as follows:

- **Section** < **Chapter** not indented
- **Section** < **Section** indented .5 inch
- **Section** < **Section** < **Section** indented 1 inch, etc.

To do this:

1. In the EDD, locate the **Section** element definition.
2. In the **Structure View**, click below **AutoInsertions** for the **Section** element.
3. From the **Elements** panel, insert **TextFormatRules**.
4. Insert **ContextRule**.

   ContextRule, If, and **Specification** elements appear.
5. In the **Specification** element, type: `Section`
6. Click below **Specification** element.
7. Specify **LeftIndentChange** of .5 in and **FirstIndentRelative** of 0
   a. Insert **ParagraphFormatting**.
   b. Insert **PropertiesBasic**.
   c. Insert **Indents**.
   d. Insert **LeftIndentChange** and type: `+.5 in`
   e. Click below **LeftIndentChange** element.
   f. Insert **FirstIndentRelative** and type: `0`
8. Save your changes.

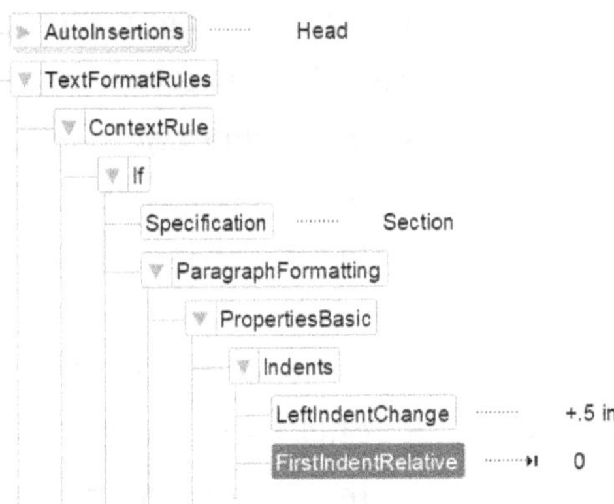

9. Reimport the element definitions from your EDD into your test document. If you get errors when importing the element definitions, correct your EDD and reimport.

   Continue to the next step when your import occurs without errors.

A nested **Section** element (and all its descendants) is progressively indented in .5 inch increments.

[[Section Head]

[Para text]

[[Second level section head]

[Para Text]

[[Third level section head]

[Para text]]]

Exercise 2: Indenting Lists

In this exercise, you will indent your **List** elements (and all their descendants) by specifying a **LeftIndentChange** of .25 inch and **FirstIndentRelative** of zero.

When using the "change" and "relative" elements, indenting of **List** elements will be added to the indenting of **Section** elements as follows:

- **List < Section < Chapter** indented .25 inch
- **List < Section < Section** indented .75 inch
- **List < Section < Section < Section** indented 1.25 inches, etc.

To indent your lists:

1. In the EDD, locate the **List** element definition.
2. In the **Structure View**, click below **AutoInsertions** for the **List** element.
3. From the **Elements** panel, insert **TextFormatRules.**
4. Insert **AllContextsRule.**
5. Specify **LeftIndentChange** of .25 in and **FirstIndentRelative** of 0
   a. Insert **ParagraphFormatting.**
   b. Insert **PropertiesBasic.**
   c. Insert **Indents.**
   d. Insert **LeftIndentChange** and type: `+.25 in`
   e. Click below **LeftIndentChange** element.
   f. Insert **FirstIndentRelative** and type: `0`
6. Save your changes.

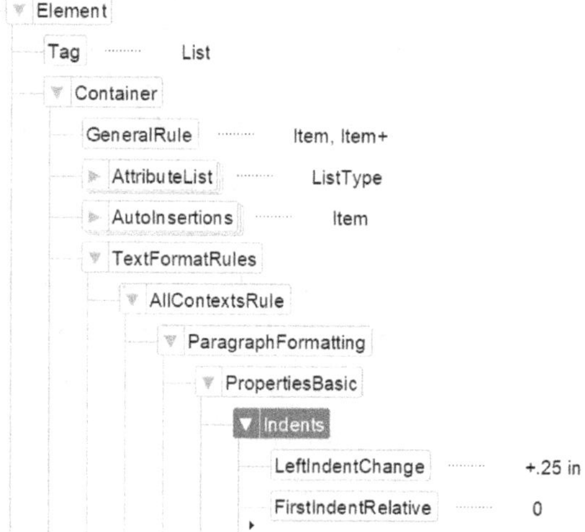

7. Reimport the element definitions from your EDD into your test document. If you get errors when importing the element definitions, correct your EDD and reimport.

   Continue to the next step when your import occurs without errors.

   The List element (and all its descendants) is indented in .25 inch more than its parent Section element.

   [[Section for a list]

   [Para describing List]

   [[[First item in list]]

   [[Second item in list]]]]

   [[Section for Table]

 Exercise 3: Specifying a ContextRule with If and Else

In this exercise, you will specify **If** and **Else** clauses for the **WarnNote**, making it bold when a child of **Section** and italic when placed elsewhere.

To specify a context rule that will do this:

1. In the EDD, locate the **WarnNote** element definition.
2. In the **Structure View**, click below the **AttributeList** element within the **WarnNote** element.
3. From the **Elements** panel, insert a **TextFormatRules** element.
4. Insert a **ContextRule** element.

   ContextRule, **If**, and **Specification** elements each appear.

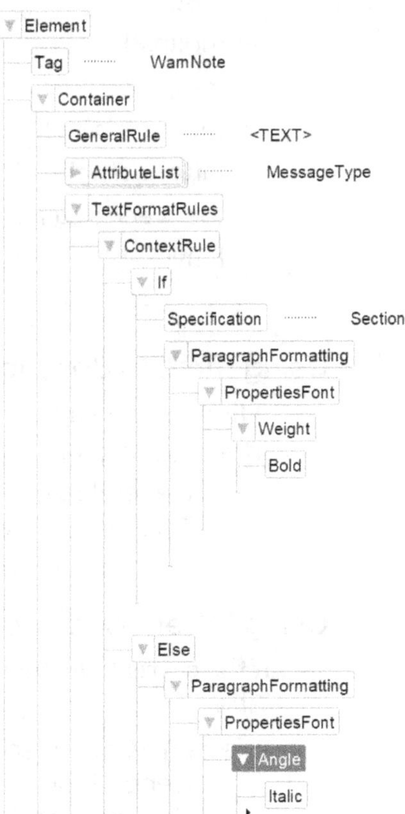

5. In the **Specification** element, type: `Section`
6. Click below **Specification** element.
7. Specify weight of **Bold**.
   a. Insert **ParagraphFormatting**.
   b. Insert **PropertiesFont**.
   c. Insert **Weight**.
   d. Insert **Bold**.
8. Click below **If** element, on line descending from **ContextRule** element.
9. Insert an **Else** element.
10. Specify **Angle** of **Italic**.
    a. Insert **ParagraphFormatting**.
    b. Insert **PropertiesFont**.
    c. Insert **Angle**.
11. Save your changes.
12. Insert **WarnNote** elements into your test document as needed to test your context rules.

13. Reimport the element definitions from your EDD into your test document. If you get errors when importing the element definitions, correct your EDD and reimport.

    Continue to the next step when your import occurs without errors.

The **WarnNote** element is bold if a child of **Section**, but is otherwise is italic.

[Para for a [*Term*] element]

[WarnNote as a child of section]

[Para text]

[[[Item text]

[WarnNote as a child of item]]

ContextRule—Naming siblings and naming attribute values

Sibling indicators describe the relationship of an element to its siblings, or an ancestor element to its siblings.

Indicator	Specification is true if element is
{first}	First element in its parent
{middle}	Neither first element nor last element in its parent
{last}	Last element in its parent
{notfirst}	Not first element in its parent
{notlast}	Not last element in its parent
{only}	Only element in its parent
{before *sibling*}	Followed by named element or text content
{after *sibling*}	Preceded by named element or text content
{between *sibling1*, *sibling2*}	Between named elements or text content
{any}	Anywhere in its parent (equivalent to no indicator)

Using {first} to describe the relationship of a current element to its siblings

- If **Item** is first child of its parent **NumberList**

```
Element (Container): Item
    General rule: <TEXT>
    Text format rules
        1.  If context is: {first} < NumberList
```

Using {after xxx} to describe a parent's relationship to its siblings

- If the current element is a child in a **Section**, only when **Section** immediately follows a sibling **Title**

```
Element (Container): Head
    General rule: <TEXT>
    Text format rules
        1.  If context is: Section {after Title} < Chapter
```

Attribute indicators describe context based on:

- Attribute name/value pair of current element
- Attribute name/value pair of ancestor element
- Set of attribute name/value pairs
- Operators with attribute name/value pairs

If/Else using an attribute and values

- **Note** formatting based on **Label** attribute value

```
Element (Container): Note
    General rule: <TEXT>
    Attribute list
        1.  Name: Label                 Choice                  Required
            Choices: Important, Note, Tip
    Text format rules
        1.  If context is: [Label = "Important"]
                Default font properties
                    Color: Red
            Else, if context is: [Label = "Note"]
                Default font properties
                    Weight: Bold
            Else
                Default font properties
                    Angle: Italic
```

If/Else using attribute and values of parent element

- If **Item** is within **List** where **Type** attribute of **List** has value of **Bullet**
- Else if **Item** is within **List** where **Type** attribute of **List** has value of **Numbered**

```
Element (Container): Item
    General rule: <TEXT>
    Text format rules
        1.  If context is: List [Type = "Bullet"]
                Numbering properties
                    Autonumber format: \b\t
                    Character format: bulletsymbol
            Else, if context is: List [Type = "Numbered"]
                Numbering properties
                    Autonumber format: <n+>\t
```

If/Else using multiple attribute and value pairs

- If **Item** is within **List** where Type attribute of **List** has value of **Num** and **Content** attribute has value of **Process**

```
Element (Container): Item
    General rule: <TEXT>
    Text format rules
        1.  If context is: List [Type="Num" & Content="Process"]
```

Attribute value comparisions with greater than and equal to

Operator	With attributes of
= (equal to)	All types
! = (not equal to)	All types
> (greater than)	Choice and numeric types
< (less than)	Choice and numeric types
>= (greater than or equal to)	Choice and numeric types
<= (less than or equal to)	Choice and numeric types

- If **Item** is within **List** where **Type** attribute of **List** does not have a value of **Numbered**

```
Element (Container): Item
    General rule: <TEXT>
    Text format rules
        1.  If context is: List [Type != "Numbered"]
```

- Operators with **Choice** attributes evaluates name/value pair using order in list of values in EDD, "lowest value" being one on left
- If **Section** is within **Report** where the **Security** attribute of **Report** is any value to the left of **Classified**

```
Element (Container): Section
    General rule: Head, Para+
    Text format rules
        1.  If context is: Report [Security < "Classified"]
```

- If **Section** is within **Report** where the **Version** attribute of **Report** has a value is between 2 and 5, inclusive

```
Element (Container): Section
    General rule: Head, Para+
    Text format rules
        1.  If context is: Report [Version >= "2" & Version <= "5"]
```

Exercise 4: Formatting Paras Using Attribute Values and Sibling Indicators

In your current structure, you have described two types of lists, **Bulleted** and **Numbered**, in the **ListType** attribute value. **List** elements contain two or more **Item** elements. **Item** elements contain one or more **Para** elements.

In this exercise, you will specify a rule for the hanging indent and tab position for the first **Para** in an **Item**, and a rule for the numbering for the first **Para** in an **Item** based on its attribute value.

In a later exercise, you'll simplify this formatting with a **SubRule**, but for now:

1. In the EDD, locate the **Para** element definition.
2. In the **Structure View**, click below the **GeneralRule** of the **Para** element.
3. From the **Elements** panel, insert **TextFormatRules**.
4. Insert **ContextRule**.
   A **ContextRule**, **If**, and **Specification** element all appear.

5. In **Specification** element, type:
   `{first} < Item`
6. Click below **Specification** element.
7. Specify **FirstIndentRelative** of `-.25 in` and **RelativeTabStopPosition** of `0`
   a. Insert **ParagraphFormatting.**
   b. Insert **PropertiesBasic.**
   c. Insert **Indents.**
   d. Insert **FirstIndentRelative** and type: `-.25 in`
   e. Click below **Indents** element.
   f. Insert **TabStops.**
   g. Insert **TabStop.**
   h. Insert **RelativeTabStopPosition** and type: `0`

Element
Tag Para
Container
GeneralRule (<TEXT> | Footnote | XRef | Inde...
TextFormatRules
ContextRule
If
Specification {first} < Item
ParagraphFormatting
PropertiesBasic
Indents
FirstIndentRelative -.25 in
TabStops
TabStop
RelativeTabStopPosition 0

8. Click below **ContextRule** element.
9. Insert another **ContextRule.**

   ContextRule, If and **Specification** elements appear.

Lessons in this chapter require typing of square brackets to invoke attributes. The specifications will be easier to type and to visualize if you hide the Element Boundaries via the View menu. However, your keyboard arrow keys will not be as effective for navigation when the Element Boundaries are not visible.

10. In **Specification** element, type:
    `{first} < Item < List [ListType = "Bulleted"]`
11. Click below **Specification** element.
12. Specify **AutonumberFormat** of `\b\t`
    a. Insert **ParagraphFormatting.**
    b. Insert **PropertiesNumbering.**
    c. Insert **AutonumberFormat.**
    d. Insert **Bullet.**
    e. Insert **Tab.**

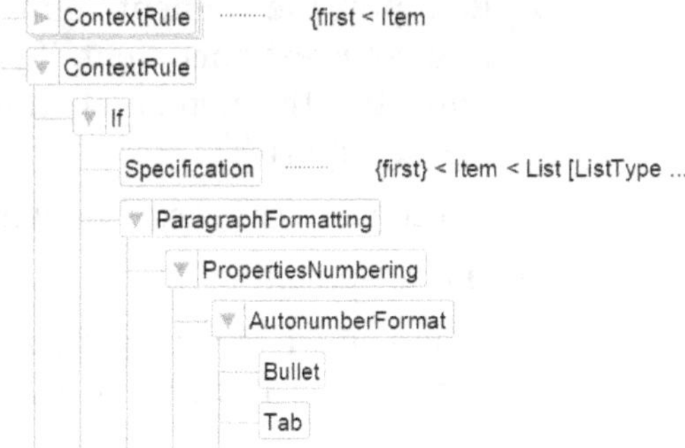

13. Collapse the **If** element and click below it.
14. Insert an **ElseIf** element.

    An **ElseIf** and a **Specification** element appear.
15. In **Specification** element, type:
    `{first} < Item {first} < List [ListType = "Numbered"]`

16. Click below the **Specification** element.
17. Specify **AutonumberFormat** of `<n=1>.\t`
    a. Insert **ParagraphFormatting**.
    b. Insert **PropertiesNumbering**.
    c. Insert **AutonumberFormat**.
    d. Insert **numberFirst**.
    e. Type a period.
    f. Insert **Tab**.

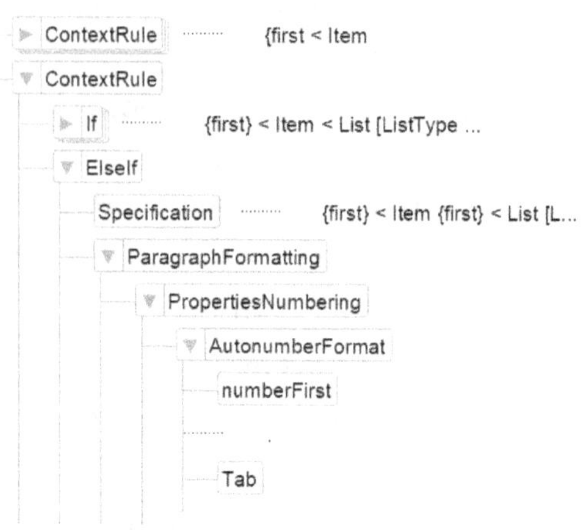

18. Collapse the **ElseIf** element and click below it.
19. Insert another **ElseIf** element.

    ElseIf and **Specification** elements appear.
20. In Specification element, type:
    `{first} < Item {notfirst} < List [ListType = "Numbered"]`
21. Click below Specification element.
22. Specify AutonumberFormat of <n+>.\t
    a. Insert **ParagraphFormatting**.
    b. Insert **PropertiesNumbering**.
    c. Insert **AutonumberFormat**.
    d. Insert **numberPlus**.
    e. Type a period.
    f. Insert **Tab**.
23. Save your changes.
24. Reimport the element definitions from your EDD into your test document. If you get errors when importing the element definitions, correct your EDD and reimport.

    Continue to the next step when your import occurs without errors.

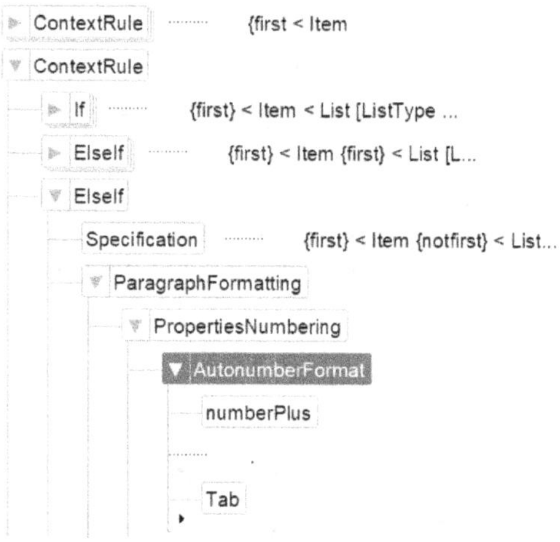

.The **Para** element:

- Has a bullet, tab and hanging indent when the first **Para** in any **Item** in a **List** of **ListType Bulleted**
- Displays with "1.", with a tab and hanging indent when the first **Para** in the first **Item** in a **List** of **ListType Numbered**
- Increments the "1." and displays with "2.", "3.", etc., with a tab and hanging indent when the first **Para** in an additional **Item** in a **List** of **ListType Numbered**

[Para before list]

1. [[[Item 1 Numbered]

   [WarnNote (child of Item)]]

2. [[Item 2 Numbered]]]

[[Section Head]

[Para before list]

- [[[Item 1 bulleted]]

Exercise 5: Controlling FrameMaker Behavior Using Attribute Values

In this exercise, you will use the attribute value of the **ReadyToImport** attribute on the **Figure** element to determine which dialog (**Import File** or **Anchored Frame**) appears when you insert **Graphic** elements. To do this:

1. In the EDD, locate the **Graphic** element definition.
2. In the **Structure View**, select the **AllContextsRule** element.
3. Delete the **AllContextsRule.**
4. Insert a **ContextRule** element.

   A **ContextRule**, an **If**, and a **Specification** element appear.
5. In the **Specification** element, type:
   `Figure [ReadyToImport = "Yes"]`
6. Click below **Specification.**
7. Insert **ImportedGraphicFile.**
8. Click below **If.**
9. Insert **Else.**
10. Insert **AnchoredFrame.**

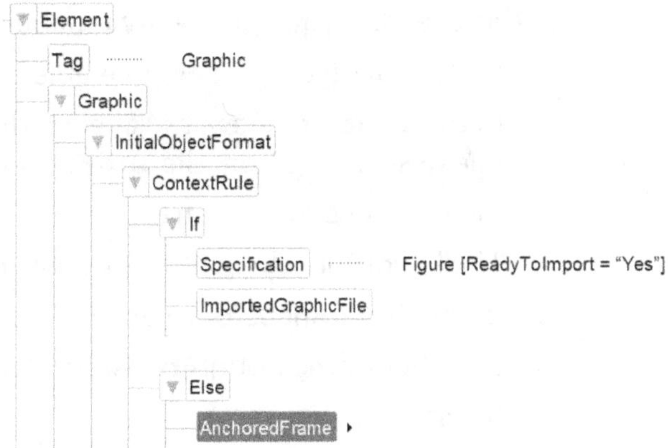

11. Save your changes.
12. Reimport the element definitions from your EDD into your test document. If you get errors when importing the element definitions, correct your EDD and reimport.

    Continue to the next step when your import occurs without errors.
13. Insert a **Figure** element with a **ReadyToImport** value of **Yes.**

    The **Import File** dialog appears.
14. Insert a **Figure** element with a **ReadyToImport** value of **No.**

    The **Anchored Frame** dialog appears.

Writing a LevelRule

Level rules define formatting changes that occur when an element is nested within a specified number of levels in an ancestor.

- In a **ContextRule**, the context of **Section < Section < Section** means "nested in at least three Section elements"
- In a **LevelRule, Section** count of 3 means "nested in exactly three **Section** elements"

Since context rules only continue processing until a condition is met, positioning a context of **Section < Section** before a context of **Section < Section < Section** would prevent the second rule ever being applied. Instead, the rules would need to be reversed, ranging from largest-to-smallest. Level rules are exact, thus can be written in a more intuitive smallest-to-largest order.

- Leve rules can also count instances of current element in hierarchy without specifying the ancestor

- Level rules can also have separate clauses for different levels using **If**, **ElseIf**, and **Else**

Each If, ElseIf, and Else clause can refer to:

- A named **ParagraphFormatTag** or **CharacterFormatTag** (if text range) stored in document
- Individual paragraph or text-range properties
- A **FormatChangeList**

Exercise 6: Numbering Headings Using LevelRules

In this exercise, you will specify numbering for **Head**, based on nesting level in **Section** ancestors.

1. In the EDD, locate the **Head** element definition.
2. In the **Structure View**, click below the **AllContextsRule** element for the **Head** element, on the line descending from the **TextFormatRules** element.
3. Insert a **LevelRule** element.

   The **LevelRule** element and **CountAncestors** child elements appear.
4. In the **CountAncestors** element, type: Section
5. Click below the **CountAncestors** element.
6. Insert an **If** element.

   The **If** element and the **Specification** child elements appear.
7. In the **Specification** element, type: 1
8. Specify **ParagraphFormatting** of C:<$chapnum>.<n+>. followed by a space
   a. Click below the **Specification** element
   b. Insert a **ParagraphFormatting** element
   c. Insert a **PropertiesNumbering** element
   d. Insert a **AutonumberFormat** element
   e. Type: C:<$chapnum>.
   f. Insert a **numberPlus** building block
   g. Type a period and space

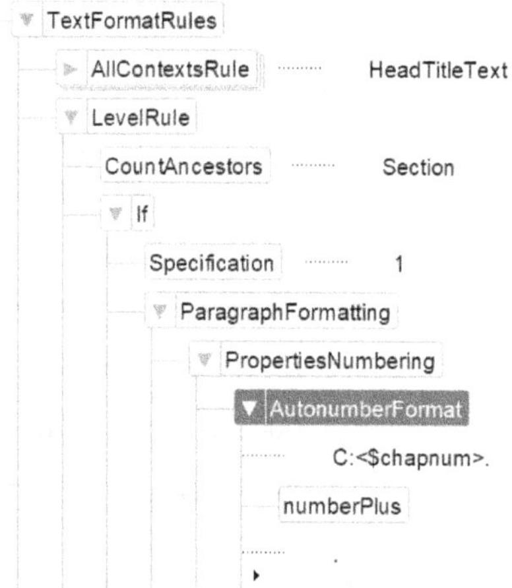

9. Click below the **If** element.
10. Insert an **ElseIf** element.

    The **ElseIf** and **Specification** child elements appear.
11. In the **Specification** element, type: 2
12. Specify **ParagraphFormatting** of C:<$chapnum>.<n>.<n+>. followed by a space
    a. Click below the **Specification** element.
    b. Insert a **ParagraphFormatting** element.
    c. Insert a **PropertiesNumbering** element.
    d. Insert a **AutonumberFormat** element.
    e. Type: C:<$chapnum>.
    f. Insert a **number** building block
    g. Type a period.
    h. Insert a **numberPlus.** building block
    i. Type a period and space.

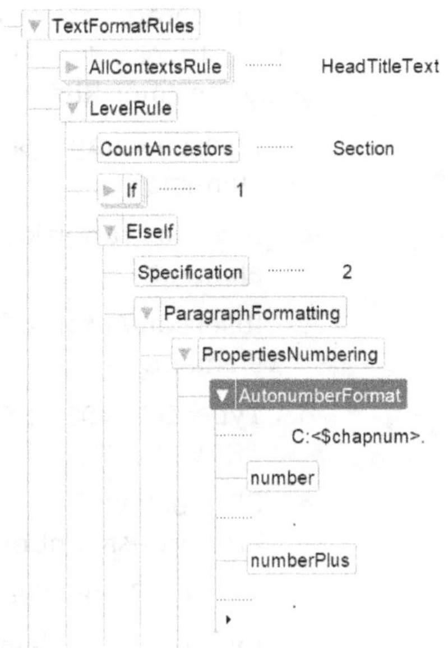

13. Click below the **ElseIf** element.
14. Insert another **ElseIf** element.

    The **ElseIf** element and a **Specification** child element appear.
15. In the **Specification** element, type: 3
16. Specify **ParagraphFormatting** of C:<$chapnum>.<n>.<n>.<n+>. followed by a space
    a. Click below the **Specification** element.
    b. Insert a **ParagraphFormatting** element.
    c. Insert a **PropertiesNumbering** element.
    d. Insert an **AutonumberFormat** element.
    e. Type: C:<$chapnum>.
    f. Insert a **number** building block.
    g. Type a period.
    h. Insert a **number** building block.
    i. Type a period.
    j. Insert a **numberPlus** building block.
    k. Type a period and space.

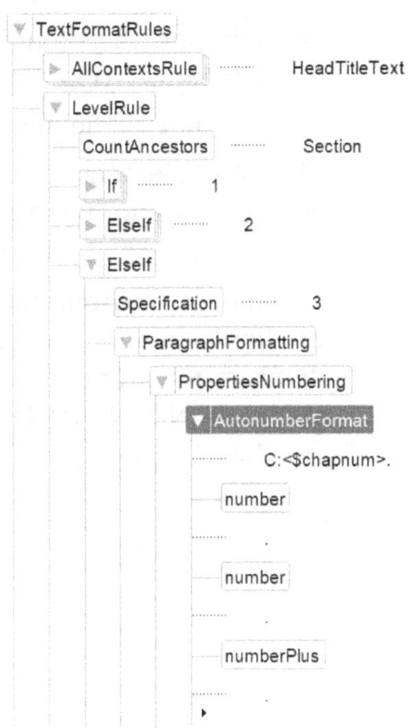

17. Click below the **ElseIf** element.
18. Insert an **Else** element.

    An **Else** element appears without a **Specification** element.

19. Specify **ParagraphFormatting** of `DO NOT INDENT TO THIS LEVEL` using the color Red.
    a. Insert a **ParagraphFormatting** element.
    b. Insert a **PropertiesNumbering** element.
    c. Insert an **AutonumberFormat** element.
    d. Type: `DO NOT INDENT TO THIS LEVEL`
    e. Click below the **PropertiesNumbering** element.
    f. Insert a **PropertiesFont** element.
    g. Insert a **Color** element.
    h. Type: `Red`

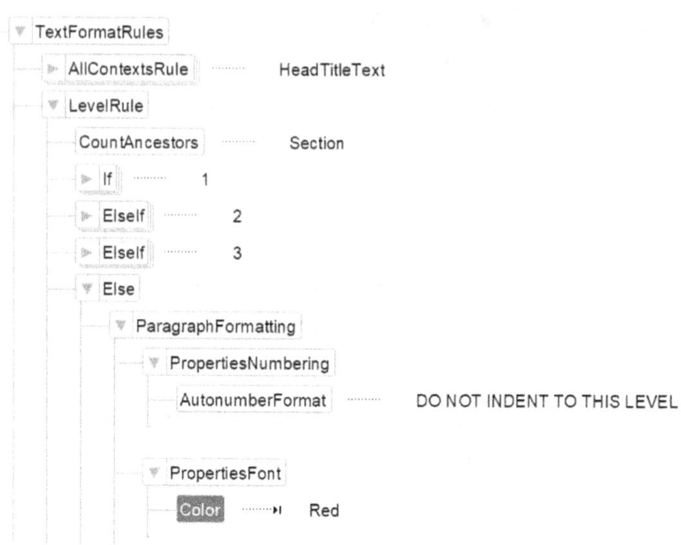

20. Save your changes.
21. Reimport the element definitions from your EDD into your test document. If you get errors when importing the element definitions, correct your EDD and reimport.

    Continue to the next step when your import occurs without errors.

The **Head** element now displays the chapter number, along with the relative nesting of the sections.

When nested within more than three **Section** elements, the **Head** element displays an autonumber, in red, of "**DO NOT INDENT TO THIS LEVEL**".

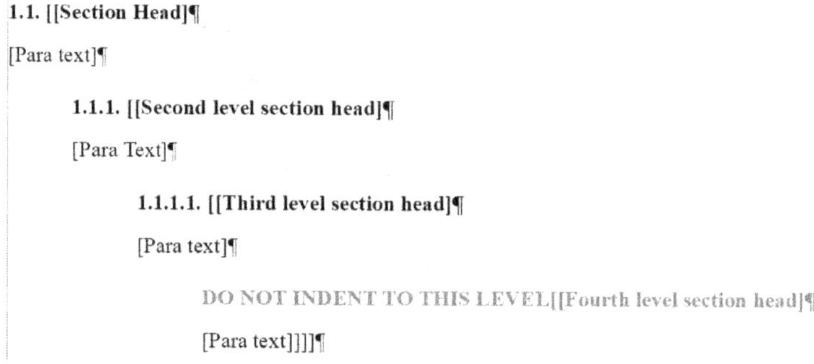

Using Context Labels

Elements can be assigned context labels in the EDD. Context labels can:

- In some dialogs, context labels display a contextual list of element tags for a user to select from

  Example: A user selects an element with a context label such as **Head (SectionHead1)** in the setup dialog for generated files such as indexes or tables of contents
- Context labels organize choices in things like the **Cross-Reference** panel, grouping elements without a label into their own list

Context labels cannot contain white-space characters or any of these special characters:

( ) & | , * + ? < > % [ ] = ! ; : { } "

Exercise 7: Providing ContextLabels for Headings

In this exercise, you will specifying context labels for **Head**, based on its nesting level in **Section** ancestors.

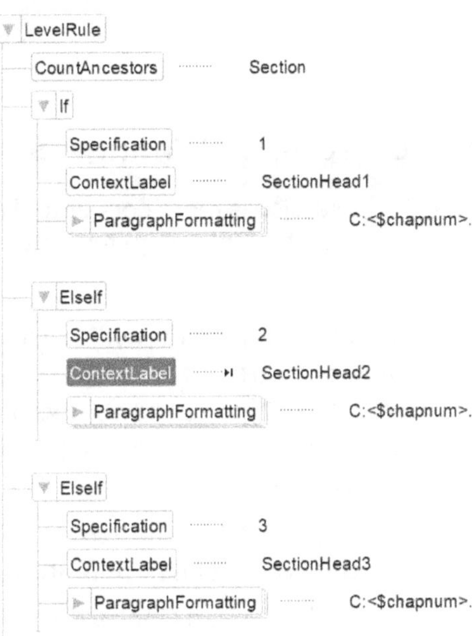

1. In the EDD, locate the **Head** element definition.
2. In the **Structure View**, click below the **Specification** element for the first **If** clause.
3. From the **Elements** panel, insert a **ContextLabel** element and label it `SectionHead1`
4. In the **Structure View**, click below the Specification element for the first ElseIf clause.
5. From the **Elements** panel, insert a **ContextLabel** element and label it `SectionHead2`
6. In the **Structure View**, click below the Specification element for the second ElseIf clause.
7. From the **Elements** panel, insert a **ContextLabel** element and label it `SectionHead3`
8. Save your changes.
9. Reimport the element definitions from your EDD into your test document. If you get errors when importing the element definitions, correct your EDD and reimport.

   Continue to the next step when your import occurs without errors.
10. Insert a Cross-Reference to a **Head** element

The **Head** element, when selected from the list of elements in something like the **Cross-Reference** panel shown below, now uses context labels to segregate levels of **Head** elements from each other:

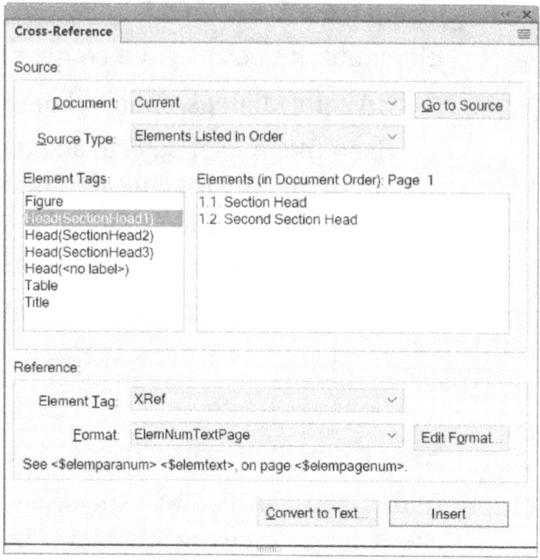

Exercise 8: Writing a SubRule for Paras in Items in Lists

This exercise is more conceptual than most. If you choose to skip it, you can use the start file specified in the next chapter to catch up to this completed exercise.

In this exercise, you will revise the **TextFormatRules** for a **Para** in an **Item** in a **List** of **ListType** of **Numbered** by replacing two **ElseIf** clauses with an **ElseIf** clause containing a **SubRule**.

1. In the EDD, in the **Structure View**, locate the **Para** element definition.
2. In the second **ContextRule** element, select the two **ElseIf** elements and delete them.

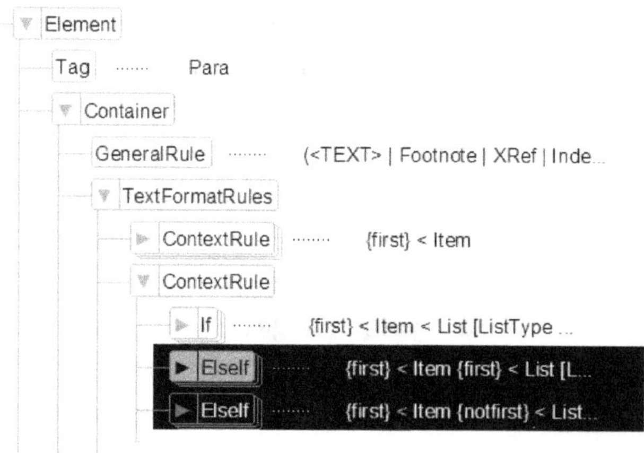

3. Insert a new **ElseIf** element in that position, with the following **Specification** (typed on a single line)

   ```
   {first} < Item <
   List[ListType="Numbered"]
   ```

4. Insert a **SubRule** element.
5. Insert a **ContextRule** element.
6. Set the **Specification** to Item {first}
7. Insert a **ParagraphFormatting** element.
8. Insert a **PropertiesNumbering** element.
9. Insert an **AutoNumber** element.
   a. Insert a **numberFirst** building block.
   b. Insert a period.
   c. Insert a **Tab** building block.
10. Insert an **Else** element after the **If** element
11. Complete the formatting as shown in the screen shot.

The **Else** statement at the end of the **SubRule** handles all first **Para** elements in **Item** elements in numbered lists that are *not* the first **Item** in the list.

12. Save your changes.
13. Reimport the element definitions from your EDD into your test document. If you get errors when importing the element definitions, correct your EDD and reimport.

    Continue to the next step when your import occurs without errors.

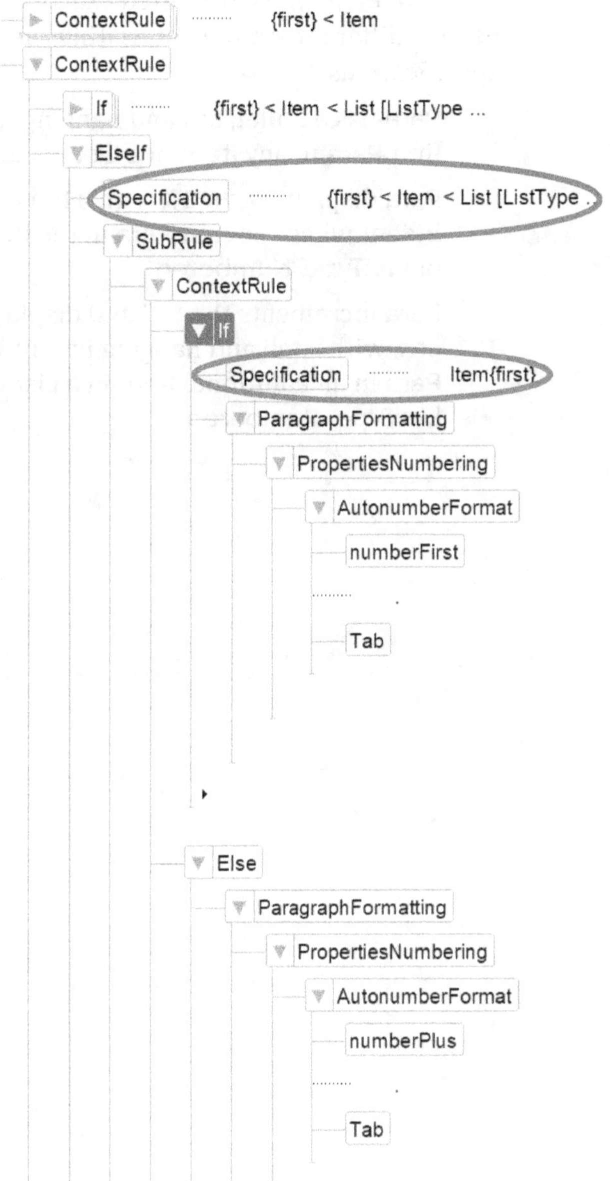

The **Para** element is formatted exactly as before, but using a different strategy of writing the formatting specifications:

- **Para** has a bullet, tab and hanging indent when the first **Para** in any **Item** in a **List** of **ListType=Bulleted**
- **Para** displays with "1.", with a tab and hanging indent when the first **Para** in the first **Item** in a **List** of **ListType=Numbered**
- **Para** increments the "1." and displays with "2.", "3.", etc., with a tab and hanging indent when the first **Para** in an additional **Item** in a **List** of **ListType=Numbered**

12. Section for lists

Para describing Numbered List

1. First item in list

    WarnNote Text

2. Second item in list

Para describing bulleted list

- First bulleted item
- Second bulleted item

Chapter 12: First/LastParagraphRules

Introduction

This chapter focuses on the formatting rules for first and last paragraphs in an element.

Objectives

- Specify **FirstParagraphRules** for formatting
- Specify **LastParagraphRules** for formatting
- Review the differences between using **First/LastParagraphRules** and specifying **{first}/{last}** sibling indicators

Overview

First/last rules are an optional part of element definitions for **Container** elements only.

- They apply a special set of format rules to the first or last paragraph in an element
- They are ignored when a first/last child element is formatted as a text range
- If an element has a prefix formatted in a separate paragraph, the prefix is the first paragraph
- If an element has a suffix formatted in a separate paragraph, the suffix is the last paragraph

You can specify **FirstParagraphRules** and **LastParagraphRules** with:

- an **AllContextsRule**
- a **ContextRule**
- a **LevelRule**

Specifying first and last paragraph rules

Exercise 1: Specifying formatting properties with first/last paragraph rules

In this exercise, you will specify that the last paragraph in a **List** element will have extra space below it, regardless of whether that paragraph is in a **WarnNote**, **Para**, or other element.

1. If it is not already open, from your class files directory, open **EDD.fm**, the EDD you'll be modifying throughout the class.

   If you did not finish the previous chapter's modifications to the EDD, please open **Chapter 12-start First Last Paras.fm** instead, and save it in your class files directory as **EDD.fm**.

For instructions on downloading class files, see "Downloading class files" on page 1.

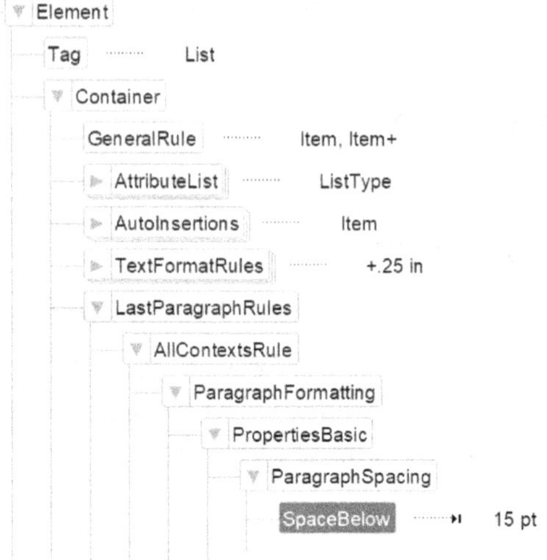

2. In the EDD, locate the **List** element definition.
3. In the **Structure View**, click below the **TextFormatRules** element.
4. From the **Elements** panel, insert a **LastParagraphRules** element.
5. Insert an **AllContextsRule** element.
6. Insert a **ParagraphFormatting** element.
7. Insert a **PropertiesBasic** element.
8. Insert a **ParagraphSpacing** element.
9. Insert a **SpaceBelow** element.
10. In the **SpaceBelow** element, type: `15 pt`
11. Save your changes.
12. Reimport the element definitions from your EDD into your test document. If you get errors when importing the element definitions, correct your EDD and reimport.

    Continue to the next step when your import occurs without errors.
13. Confirm that the last paragraph in the **List** element, regardless if it is a descendant **Para** element or **WarnNote** element, has 15 points of space below.

    If you had specified **TextFormatRules** of **SpaceBelow** 15 points for the **List** element, rather than **LastParagraphRules**, each descendant **Item** and their children would have 15 points below, unless you wrote another rule counteracting that inheritance.

    If you had used **{last}** sibling indicators, you would have had to define the **{last}** sibling indicators for all potential last children.

1.1. Section Head

Para before list

• Item 1 bulleted

• Item 1 bulleted

Para below List

If you want to see the actual values applied, display the Paragraph Designer and view the Basic properties when you select the various parts of the list.

12. Section for lists

Para describing Numbered List

1. First item in list

   WarnNote Text

2. Second item in list

Para describing bulleted list

• First bulleted item

• Second bulleted item

13. Section for Table

Chapter 13: PrefixRules and SuffixRules

Introduction

This chapter focuses on defining the **PrefixRules** and **SuffixRules** for **Container** elements.

Objectives

- Define prefixes and suffixes
- Specify **PrefixRules** and **SuffixRules** using fixed text string
- Specify **PrefixRules** and **SuffixRules** referring to attribute values
- Specify formatting for a prefix and a suffix
- Compare use of autonumbers, **Prefix/SuffixRules**, and **First/LastParagraphRules**

Overview

Prefix and suffix are an optional part of element definitions for **Container** elements only

- A prefix is a text range defined in the EDD that appears at beginning of element (before element's content)
- A suffix is a text range defined in the EDD that appears at end of element (after content)
- **PrefixRules** and **SuffixRules** describe both text string and any special font properties
- Format rules for the prefix/suffix describe font changes only for the prefix/suffix

  Font changes do not apply to descendants
- If an element has **FirstParagraphRules** and **LastParagraphRules** and a prefix/suffix is formatted in a paragraph of its own, the prefix/suffix is first/last paragraph for formatting
- Because **PrefixRules** and **SuffixRules** are applied after first/last rules, **PrefixRules** and **SuffixRules** can override font changes in a first/last rule

Examples:

Using a Prefix/suffix to wrap a text range element in the middle of a paragraph

- Result: Display double quotation marks around the text of a quotation

Element (Container): Quotation
General rule: <TEXT>
Text format rules
1. In all contexts.
Text range.
Prefix rules
1. In all contexts.
Prefix: "
Suffix rules
1. In all contexts.
Suffix: "

Using a Prefix/suffix to label a paragraph, similar to an autonumber

- Result: Display **Important:** at beginning of paragraph

Element (Container): Note
General rule: <TEXT>
Prefix rules
1. In all contexts.
Prefix: Important:

Prefix/suffix for an element with a sequence of paragraphs

- Result: Display a bold string like **Synopsis and Contents**, **Arguments**, or **Examples** in a paragraph by itself above a first child of **Para**

Element (Container): Syntax
General rule: Para+
Prefix rules
1. If context is: Synopsis
Prefix: Synopsis and Contents
Else, if context is: Args
Prefix: Arguments
Else, if context is: Examples
Prefix: Examples
2. In all contexts.
Text range.
Font properties
Weight: Bold
Format rules for first paragraph in element
1. In all contexts
Basic properties
Paragraph spacing
Space below: 4pt

The format rules for the first paragraph puts 4 points of space below the prefix paragraph

Prefix/suffix for both text range and paragraph

- Result: Display an **AuthorNote** element within a **Para** as text range, and elsewhere as a paragraph, with prefix/suffix

```
Element (Container): AuthorNote
    General rule: <TEXT>
    Text format rules
        1.  If context is: Para
                Text range.
                Font properties
                    Angle: Italic
            Else
                Default font properties
                    Angle: Italic
    Prefix rules
        1.  In all contexts.
                Prefix: [Author's comments:
                Text range.
                    Font properties
                            Weight: Bold
    Suffix rules
        1.  In all contexts.
                Suffix: ]
                Text range.
                    Font properties
                            Weight: Bold
```

Displaying attribute values using PrefixRules and SuffixRules

- Result: Display value of **Security** attribute in current element

  Prefix: <$attribute[Security]>

- Result: Display value of **Security** attribute in closest ancestor **Item** or **LabelPara**

  Prefix: <$attribute[Security: Item, LabelPara]>

- Result: Display value of **Security** attribute in closest ancestor **Head** element with **Chapter Level** context label

  Prefix: <$attribute[Security: Head(Chapter Level)]>

Specifying PrefixRules and SuffixRules

Exercise 1: Displaying an attribute value as a prefix

In this exercise, you will use the **MessageType** attribute value of a **WarnNote** element as a prefix for the element and apply to it the color **Red**.

1. If it is not already open, from your class files directory, open **EDD.fm**, the EDD you'll be modifying throughout the class.

   If you did not finish the previous chapter's modifications to the EDD, please open **Chapter 13-start Prefix Suffix.fm** instead, and save it in your class files directory as **EDD.fm**.

For instructions on downloading class files, see "Downloading class files" on page 1.

2. In the EDD, locate the **WarnNote** element definition.
3. In the **Structure View**, click below the **TextFormatRules** element.
4. From the **Elements** panel, insert a **PrefixRules** element.
5. Insert a **AllContextsRule** element.
6. Insert a **Prefix** element.
7. Insert an **AttributeValue** element.
8. In the **AttributeValue** element, type: `MessageType`
9. Click below the **AttributeValue** element.
10. Type a colon (:) followed by a space.
11. Click below the **Prefix** element.
12. Insert a **TextRangeFormatting** element.

    The **TextRangeFormatting** and **TextRange** elements appear.
13. Insert a **PropertiesFont** element.
14. Insert a **Color** element.
15. In the **Color** element, type: `Red`
16. Save your changes.
17. Reimport the element definitions from your EDD into your test document. If you get errors when importing the element definitions, correct your EDD and reimport.

    Continue to the next step when your import occurs without errors.

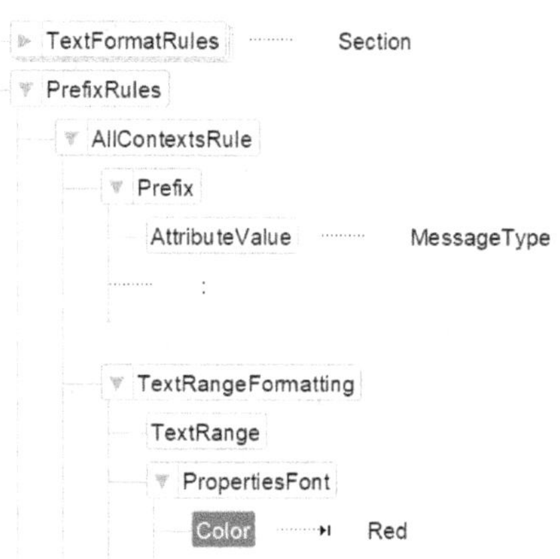

The **WarnNote** element displays a prefix (in red) of:

Para for a *Term* element

WARNING: **WarnNote with MessageType=WARNING**

Para text

NOTE: **WarnNote with MessageType=NOTE**

Para text

- "**NOTE:** " if using **MessageType** attribute value NOTE.
- "**WARNING:** " if using **MessageType** attribute value WARNING.

Exercise 2: Displaying Quotation Marks Around a Text-Range Element

In this exercise, you will use both a **PrefixRule** and a **SuffixRule** to add quotation marks around the **Term** element.

1. In the EDD, locate the **Term** element definition.
2. In the **Structure View**, click below the **TextFormatRules** for the **Term** element.
3. Insert a **PrefixRules** element.
4. Attempt the rest on your own, referring to the picture below.
5. Save your changes.
6. Reimport the element definitions from your EDD into your test document. If you get errors when importing the element definitions, correct your EDD and reimport.

   Continue to the next step when your import occurs without errors.

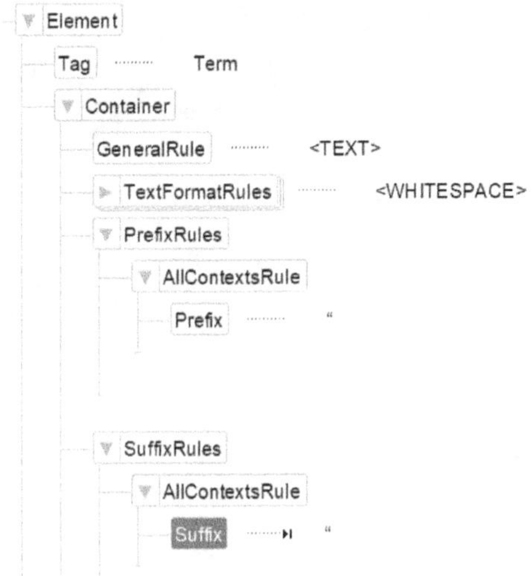

The **Term** element has curved quotes around it in the Document View.

1.1.2. Section Head

Para for a "*Term*" element

Based on your system, you may want to specify straight quotes, or specific curved quote characters with FrameMaker character codes. See *FrameMaker - Working with Content* for details.

Chapter 14: Elements for Structuring Books

Introduction

In this chapter, you will define book-related elements in the EDD, import them into a chapter in a book, generate the book, adding additional chapters, and a table of contents and index.

Objectives

- Define an element as a book
- Define elements for generated files within a book
- Generate a structured book
- Add files to a structured book
- Generate and format a table of contents
- Generate and format an index
- Wrap generated files into elements of a structured book

Overview

In FrameMaker you can use books to:

- Maintain several documents as one larger document
- Generate table of contents, index, list of figures, list of tables for several files at once
- Allow page sides, page numbering, paragraph numbering to continue across files
- Open, save, print and close all files at once
- Enforce consistent structure of all files within book

A structured book has its own:

- Element hierarchy, which display in the **Structure View**
- List of elements, which display in the **Elements** panel

A book's element definitions are defined in same EDD shared by the book's chapters

- They define a container element for the book and specify that it is a **ValidHighestLevel** element
- They define a container element for each structured file within the book and specify those elements as **ValidHighestLevel**
- They define a container element for each generated (unstructured) file within the book with a **GeneralRule** of **<TEXT>**

Defining and Testing Book Elements

Exercise 1: Defining an Element for the Entire Book

In this exercise, you will define a **Container** element that is a **ValidHighestLevel** element for the book as a whole.

1. If it is not already open, from your class files directory, open **EDD.fm**, the EDD you'll be modifying throughout the class.

   If you did not finish the previous chapter's modifications to the EDD, please open **Chapter 14-start Book Elements.fm** instead, and save it in your class files directory as **EDD.fm**.

For instructions on downloading class files, see "Downloading class files" on page 1.

2. In the EDD, click on the line descending from the **ElementCatalog** element, anywhere above the **FormatChangeList** element at the very end.
3. Create a new **Element** and tag it **UserManual**.
   a. From the **Elements** panel, insert an **Element** element.

      An **Element** and a child **Tag** element appear.

   b. In the **Tag** element, type: `UserManual`
4. Define **UserManual** as a **Container** with a **GeneralRule** of `TOC, Chapter, Chapter+, IX?`
   a. Click below the **Tag** element.
   b. From the **Elements** panel, insert a **Container** element.

      The **Container** element and a **GeneralRule** child element appear.

   c. In the **GeneralRule** element, type: `TOC, Chapter, Chapter+, IX?`
5. Define the **UserManual** element as a **ValidHighestLevel** element.

   a. Click above or below the **GeneralRule** element, on the line descending from the **Container** element.
   b. From the **Elements** panel, insert **ValidHighestLevel**.

      ValidHighestLevel element and **Yes** child element appear.
6. Save your changes.

Exercise 2: Defining Elements for the Generated Files

You specified two new elements in the previous exercise so in this exercise you will define **Container** elements for these two generated file elements—**TOC** and **IX**.

1. Create a new **Element** and tag it **TOC**.
   a. From the **Elements** panel, insert an **Element** element.
      An **Element** and a child **Tag** element appear.
   b. In the **Tag** element, type: `TOC`
2. Define **TOC** as a **Container** with a **GeneralRule** of <TEXT>
   a. Click below the **Tag** element.
   b. From the **Elements** panel, insert **Container**.
      Container element and **GeneralRule** child element appear.
   c. In the **GeneralRule** element, type: `<TEXT>`
3. Create a new **Element** and tag it `IX`.
   a. From the **Elements** panel, insert an **Element** element.
      An **Element** and a child **Tag** element appear.
   b. In the **Tag** element, type: `IX`
4. Define **IX** as a **Container** with a **GeneralRule** of <TEXT>
   a. Click below the **Tag** element.
   b. From the **Elements** panel, insert **Container**.
      Container element and **GeneralRule** child element appear.
   c. In the **GeneralRule** element, type: **`<TEXT>`**
5. Save your changes.

Exercise 3: Reimporting and Retesting

In this exercise, you will reimport the EDD into the structured template and test your element definitions for **UserManual**, **TOC**, and **IX**.

> Although you will not be using these elements in a single file, it is best to test them in the structured template before importing into the book.

1. Reimport the element definitions from your EDD into your test document. If you get errors when importing the element definitions, correct your EDD and reimport.
   Continue to the next step when your import occurs without errors.

2. In the **Structure View**, select the **Chapter** element (at the top of the structure) and delete it.

   The **Elements** panel displays the **UserManual** element as a valid element to enter into this document.

3. Insert a **UserManual** element.

   The **UserManual** is inserted, and the **Elements** panel displays the **TOC** element as valid at this position.

4. Insert a child **TOC** element.

   The **Elements** panel displays **<TEXT>** as valid input in the **TOC**.

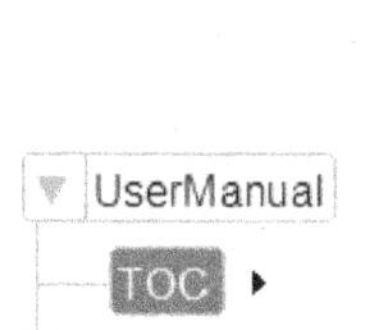

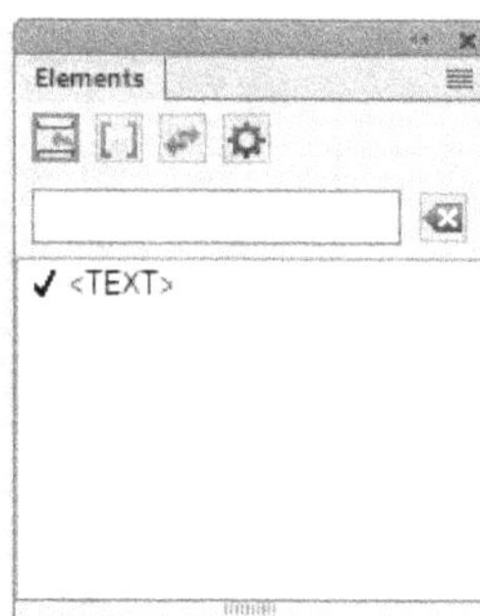

5. In the **Structure View**, click below the **TOC** element.

   The **Elements** panel displays the **Chapter** element as valid.

6. Insert **Chapter** then click below it in the **Structure View**.

   The **Elements** panel displays the **Chapter** element as valid again (you need a minimum of two).

7. Insert **Chapter** then click below it in the **Structure View**.

   The **Elements** panel displays the **Chapter** and **IX** elements as valid.

8. Insert **IX**.

   The **Elements** panel displays **<TEXT>** as valid.

9. Save your changes.

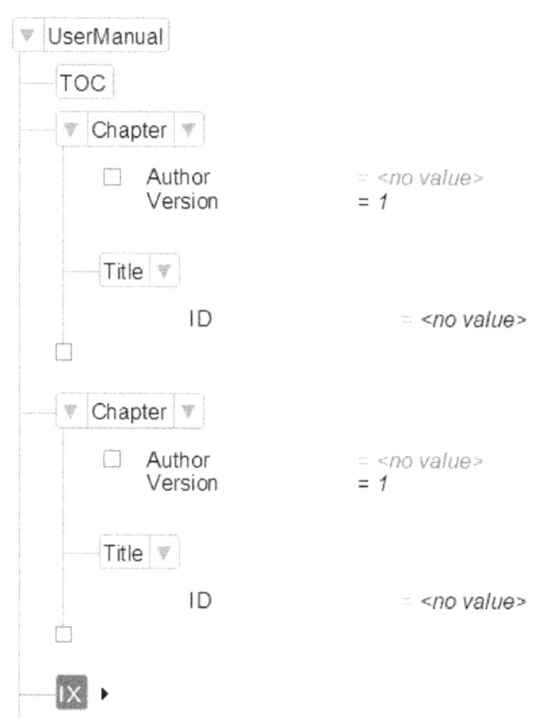

While this structure looks like a book, it isn't as useful as the FrameMaker book structure you may be used to and the numbering isn't acting as we would expect.

In the rest of this lesson you will create a "normal" FrameMaker book with a generated TOC and Index.

Generating a FrameMaker Book

Exercise 4: Importing the Element Definitions into sample content

In this exercise, you will import your element definitions into a chapter from which you will then generate the book. The chapter is already structured, using the version of the EDD without the book definitions. Because you are importing into the chapter before generating the book, the book will automatically have the book's element definitions available in its element catalog.

1. From your class files directory, open **chap1.fm**.
   a. From the **File** menu, choose **Open**.

      The **Open** dialog appears.
   b. If necessary, change to your class files directory.
   c. Double-click **chap1.fm**.

      The sample structured document appears. This document already contains the full EDD needed to complete the exercises.
2. Save your changes.

Exercise 5: Generating a Book

In this exercise, you will create and save a book.

1. With your cursor in the **chap1.fm** file, choose **File > New > Book** from the **File** menu.

   An alert box appears asking "**Do you want to add the file'chap1.fm' to the new book?**"

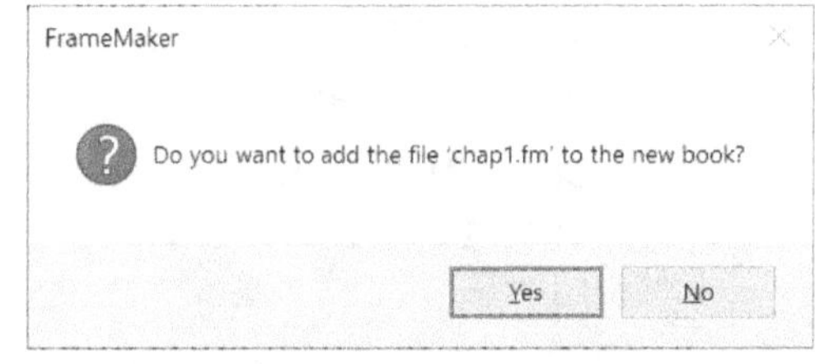

2. Select the **Yes** button to proceed.

   A new book window appears.

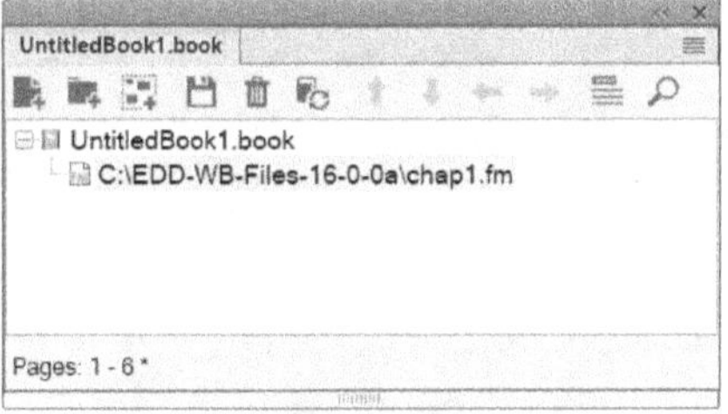

Notice that FrameMaker shows the path of the **chap1.fm** file, but not the path of the book. You will save the book to your disk. Afterward, the path of the book will show, and only a relative path to the **chap1.fm** file will remain.

3. With the book window active, choose **File> Save Book As.**
4. If needed, navigate to your class file directory and in the **File name** text box, replace the contents with: `UserManual.book`

You can manually type in the **.book** extension, but FrameMaker will add it for you if you forget.

5. Click **Save**.

   The book is now saved with its new name.

   Note the lack of a path for **chap1.fm**, as it is in the same directory as **UserManual.book**.

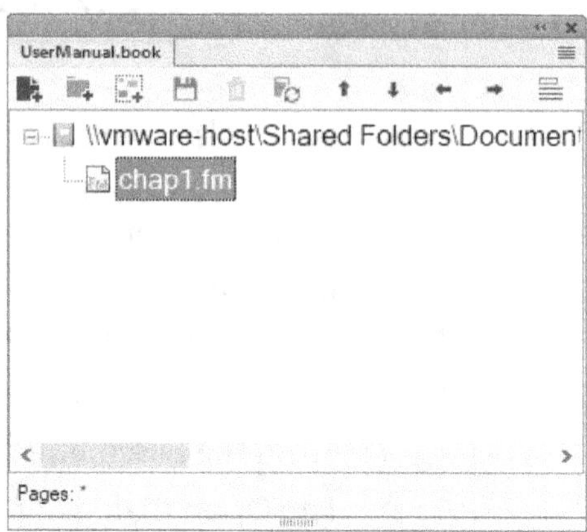

Exercise 6: Adding Files

In this exercise, you will add three more chapters to the book.

1. With the book window active, choose **Insert>Files**.

   The **Add Files to Book** dialog appears.

2. Select the **chap2.fm**, **chap3.fm** and **chap4.fm** files.

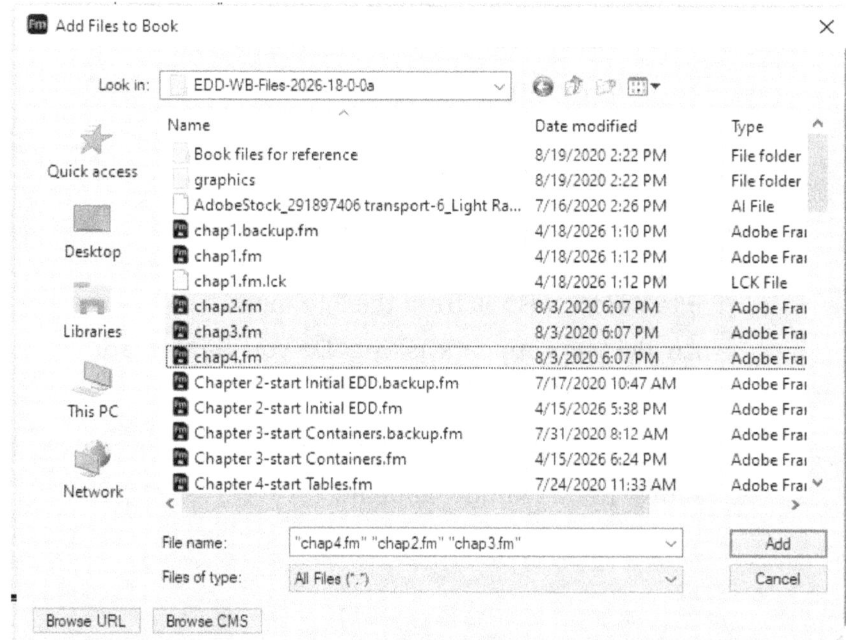

3. Click **Add**.

   The book window updates to display the added files. You'll rearrange the files in the next exercise.

4. With the book window active, from the **File** menu, choose **Save**.

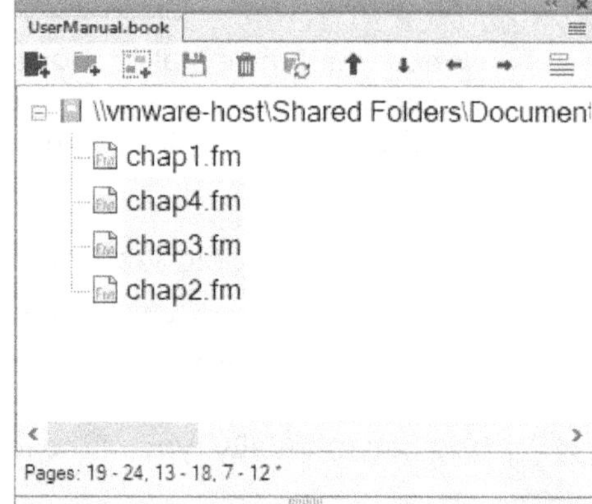

Exercise 7: Rearranging Files in the Book

In this exercise, you will rearrange the files in the book, putting the four chapters in numerical order.

If your chapters are already in numerical order, skip to the next exercise.

1. Rearrange your files as needed by selecting and dragging individual files in either the book window or the structure window.

   You can also move their position by using the arrows above the book name.

2. Move all the files in the book until they are all in the correct order.

3. From the **File** menu, choose **Save Book**.

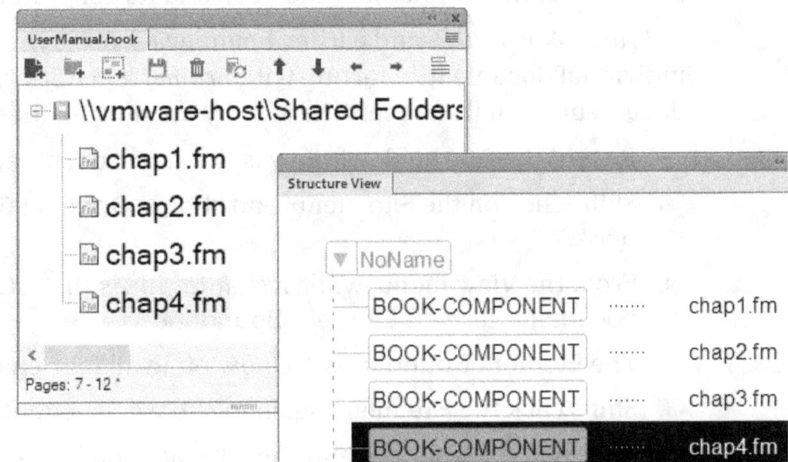

The **Structure View** displays the label for the highest-level element as **NoName** and each file is labeled as **BOOK-COMPONENT**. You'll correct this in the next exercise.

Exercise 8: Correcting a NoName Element

The **NoName** element at the top of the book window indicates that FrameMaker doesn't yet know what part of the structure should be applied here.

In this exercise, you will change the **NoName** element into a **UserManual** element.

1. In the **Structure View**, select the **NoName** element.

2. From the **Elements** panel, select **UserManual** and click the **Change button** ( ).

   The **Structure View** now shows the highest-level element as **UserManual**, without displaying its label in red.

3. Save your changes.

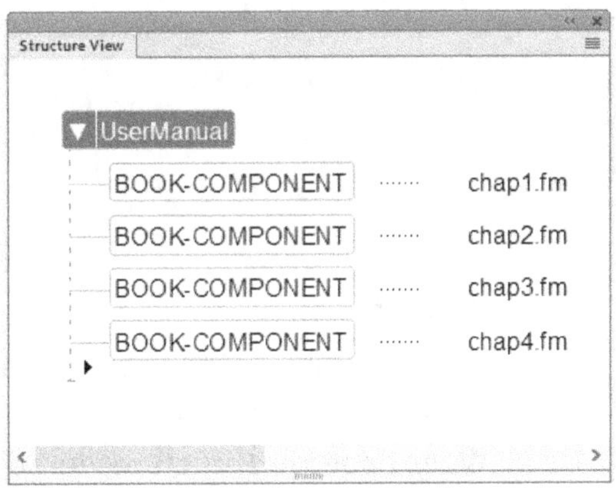

Exercise 9: Updating the book to correct Chapter elements in Structure View

In this exercise, you will update the book. Before you update the book, you need to turn off element boundaries in each file. Element boundaries take up space and, therefore, affect the page breaks and page numbering throughout the book and its generated table of contents and index.

In *FrameMaker 2015* and earlier, boundaries, text symbols, and other items were managed by individual documents. Starting with *FrameMaker 2017*, these settings are changed for all open documents. You'll start this exercise by opening all files in book to take advantage of this feature.

1. Ensure that the book window is the active document.
2. Shift+click on the **File** menu, and choose **Open all files in book.**

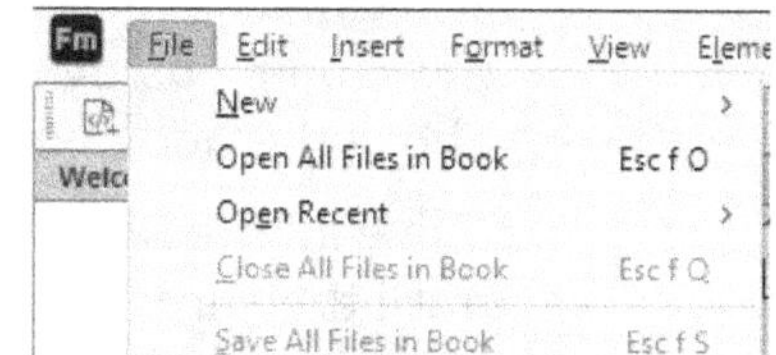

3. From the **View** menu, while a chapter file is the active document, turn off **Element Boundaries.**

   The **Element Boundaries** are now off in all four chapter files.
4. Shift+click the **File** menu again to choose **Save All Open Files** or **Save All Files in Book.**

   The choice you see depends on whether you had a chapter file or the book file active when you Shift+clicked the **File** menu.
5. With the book window active, choose **Edit>Update Book**, or use the **Update** button ( ) in the book file.

   The **Update Book** dialog appears.

   Because you have not yet added any placeholders for generated files, nothing appears in the **Generate** and **Don't Generate** fields.

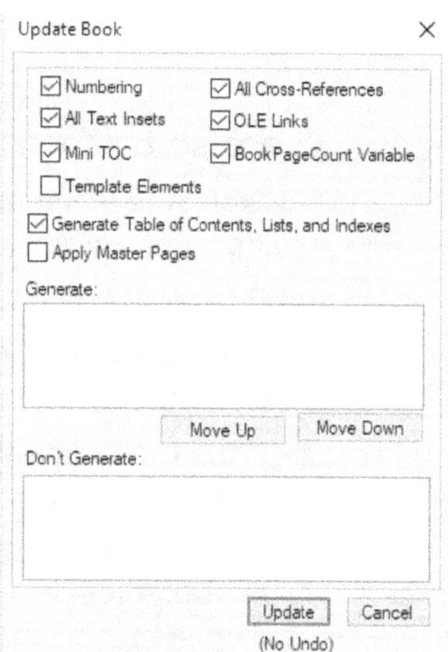

6. Click **Update.**

   Messages appear at the bottom of the book window showing the progress of the update.

   When done, the **Structure View** displays each file as **Chapter**, the highest-level element in each file.

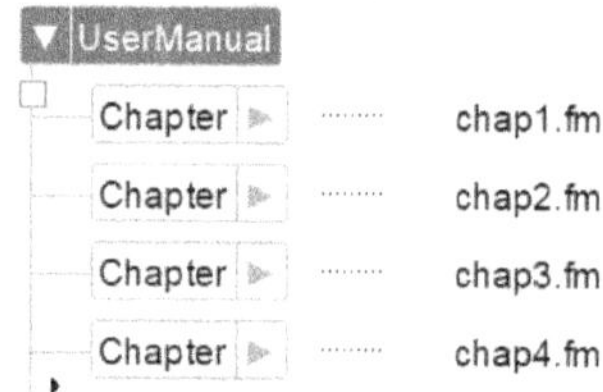

7. With the book window active, shift+click the **File** menu, and choose **Save All Files in Book.**

Exercise 10: Adding a Table of Contents

In this exercise, you will add a table of contents to the book.

In the book window, each generated file will have an orange icon to indicate that it is a generated file.

1. In the **Structure View** for your book, position your cursor to the right of the missing content. (to the right of the red square)

2. From the **Insert** menu, choose **Create TOC.**

   The **Setup Table of Contents** dialog appears.

   Because you positioned your cursor above **chap1.fm**, the default settings will place the TOC accurately. Otherwise, you might need to reposition the TOC in the future and regenerate for expected results.

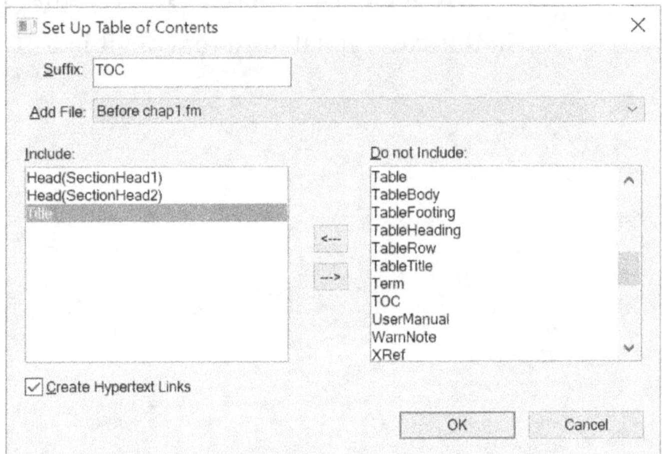

3. Move the following elements to the **Include** scroll list:
   - Head (SectionHead1)
   - Head (SectionHead2)
   - Title
4. Turn on **Create Hypertext Links.**

Hypertext links will be added to the table of contents, making the TOC clickable in PDF and other formats.

5. Click **OK** to bring up the **Update Book** dialog.
6. Click **Update** to process the **TOC** for the book.

   Dismiss any **Book Error** processing messages. Later you will import element and formatting definitions to clear these up.

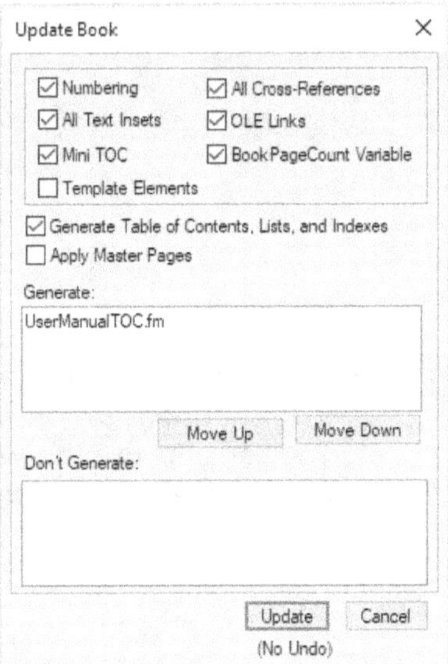

The **TOC** has been added to the book, but is represented by the BOOK COMPONENT element in the **Structure View**.

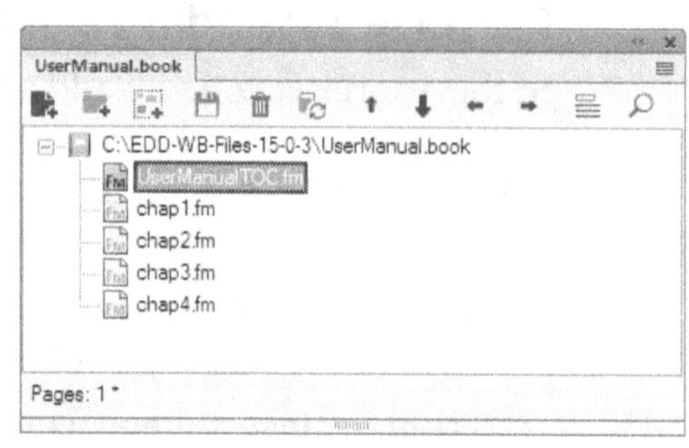

This unstructured FrameMaker document needs to be wrapped in an appropriate structured element.

In this case, you'll need to wrap it in a **TOC** element.

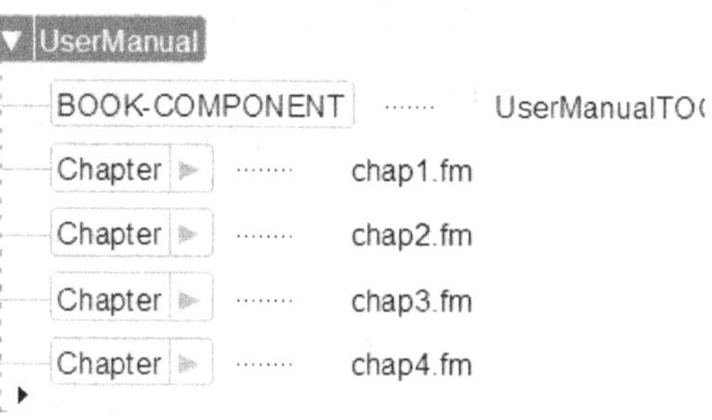

7. Select the **BOOK COMPONENT** in the book file **Structure View** and select the Wrap button ( ) in the **Elements** panel to wrap it in a **TOC** element.

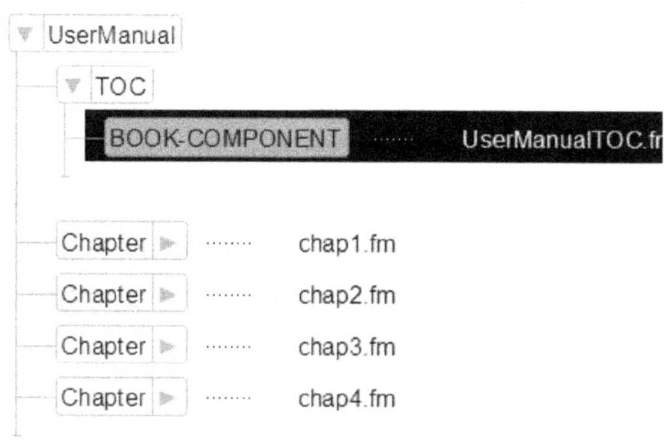

Prior to FrameMaker v18.0.0, FrameMaker had a long-standing behavior of opening new generated files like TOC and IX into their own window, rather than opening into the tabbed interface.
If this happens to you, just drag the tab of your generated file into the other tabs for your project.
You can also use the **Window > Consolidate** command to collect files into the standard tabbed interface. After opening them for the first time, the files will open into tabs as expected.

Exercise 11: Adding an index

Your sample documents already contain index markers, and in this exercise you will collect them into a generated index file in your book.

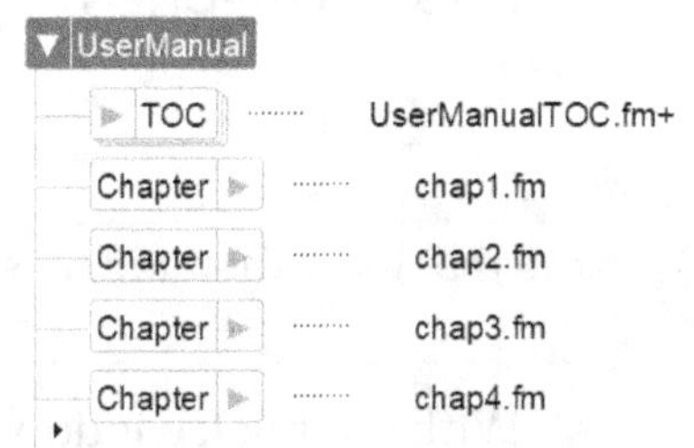

1. In the structure window, place your cursor at the end of the book.

2. Choose **Insert > Standard Index.** Change your settings to match the image, if needed.

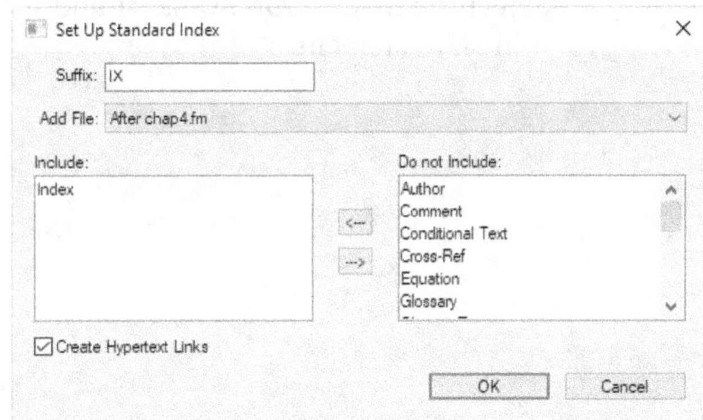

3. Click **OK.**

   The **Update Book** dialog appears.

   Make sure that both the **TOC** and **IX** files are in the **Generate** column.

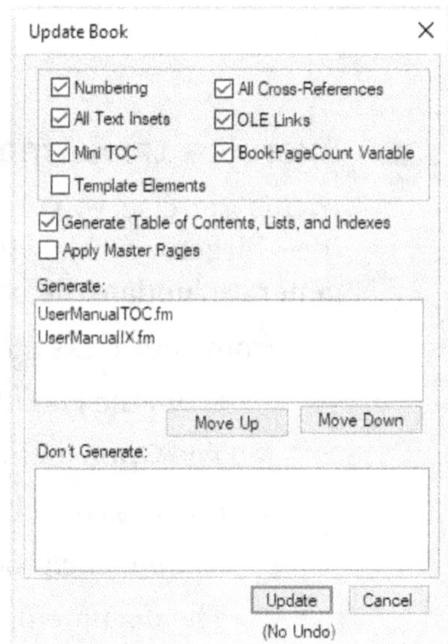

4. Click **Update.**

   The book window updates to display a placeholder for the generated index.

   Notice the difference in color between the generated and non-generated files.

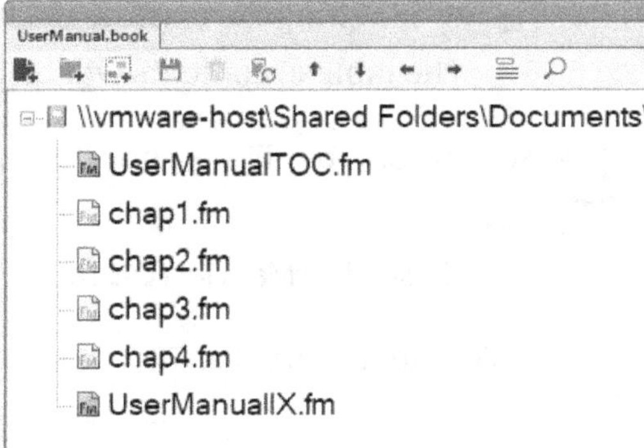

The **Structure View** displays the index as **BOOK-COMPONENT.**

Note that the generated files have a + suffix applied, further helping to identify them as generated files.

5. With the book window active, from the **File** menu, choose **Save Book.**
6. Select the **BOOK-COMPONENT** and wrap it in an **IX** element.

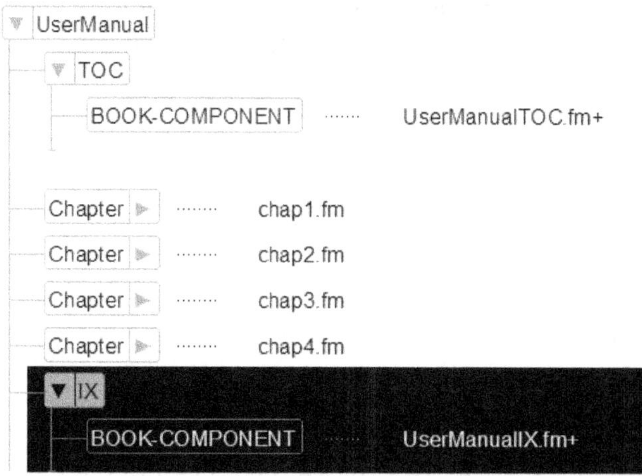

Exercise 12: Formatting a Table of Contents

In this exercise, you will open a table of contents template and save it with a new name in the book file's directory. This will result in an automatically formatted table of contents when you generate/update the book.

1. From your class files directory, open **toc.tpl.fm.**
   a. From the **File** menu, choose **Open.**

      The **Open** dialog appears.
   b. If necessary, change to your class files directory.
   c. Double-click **toc.tpl.fm.**

      The document appears.
2. Choose **File>Save As**, save the file as **UserManualTOC.fm** in your class files directory, overwriting the existing file if prompted.

   The table of contents will be populated when you update the book in a later exercise.

You can also use **File > Import > Formats** to import toc.tpl.fm formats into the UserManualTOC.fm file.

3. Close **UserManualTOC.fm**

Make sure to close the TOC so that the Update Book exercise later works as expected.

To learn more about formatting tables of contents, indexes, and other FrameMaker formatting options, see my reference book, *FrameMaker - Working with Content*, or consider taking a template design course. Information on both books and courses is available at www.techcommtools.com.

Exercise 13: Formatting an Index

In this exercise, you will open an index template and save it with a new name in the book file's directory.

1. From your class files directory, open **ix.tpl.fm**.
   a. From the **File** menu, choose **Open**.

      The **Open** dialog appears.
   b. If necessary, change to your class files directory.
   c. Double-click **ix.tpl.fm**.

      The document appears.
2. Using Save As, save **ix.tpl** as **UserManualIX.fm** in your class files directory, overwriting the existing file if prompted.

   The index will be populated when you generate/update the book in the next exercise.

You can also use **File > Import > Formats** to import the IX.tpl.fm formats into the UserManualIX.fm file.

3. Close **UserManualIX.fm**.

Make sure to close the IX so that the Update Book exercise later works as expected.

Exercise 14: Setting up numbering and pagination in a book file

In this exercise, you will set up the page side, page numbering, paragraph numbering, and prefix for each file in the book.

1. In the book window, right-click on **UserManualTOC.fm** and choose **Numbering**.

   The **Numbering Properties** dialog appears.
2. On the **Page** tab, match the **First Page #** and **Format** options as shown.
3. Click **Set**.

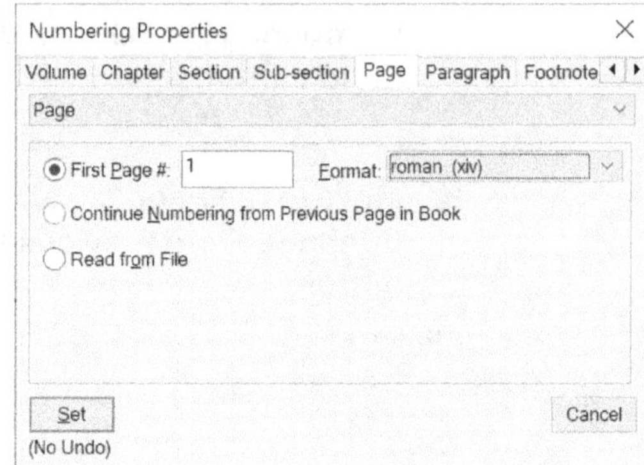

4. In the book window, right-click on **chap1.fm** and choose **Numbering.**

    a. On the **Page** tab, set **First Page** to 1 and format to **Numeric.**

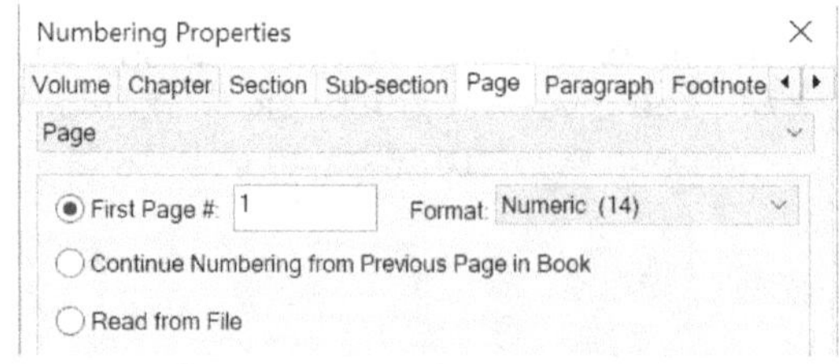

    b. On the **Chapter** tab set the properties as shown.

    c. Click **Set.**

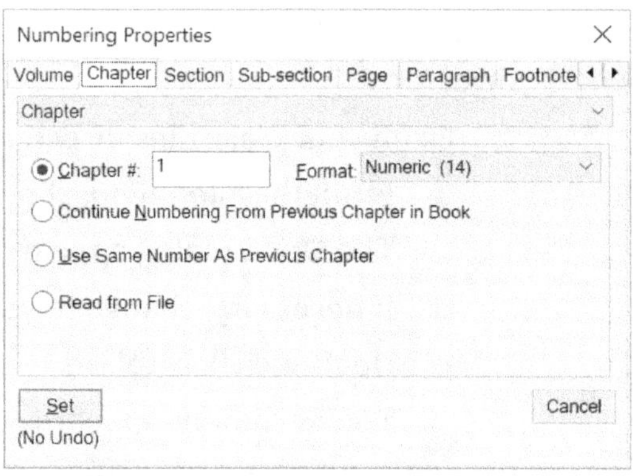

5. Select all the files after **chap1.fm** in the book window.

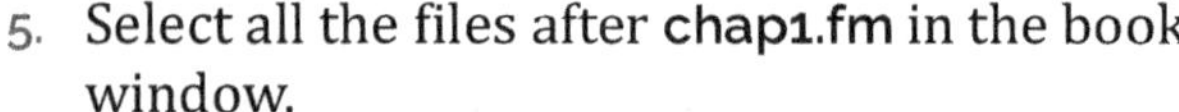

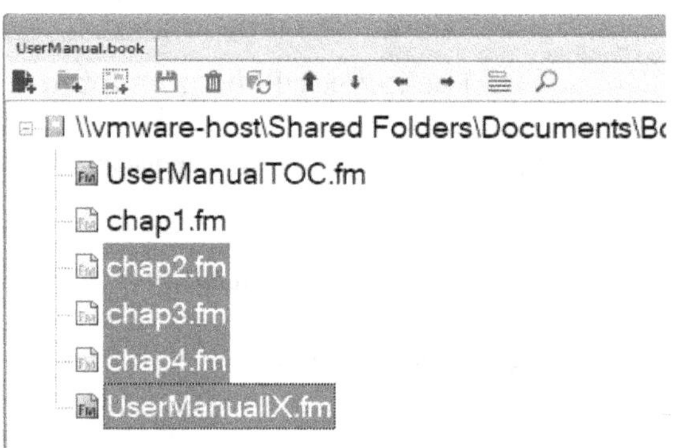

6. Right-click on the selected files and choose **Numbering.**

    a. On the **Chapter** tab, select the **Continue Numbering From Previous Chapter in Book** radio button.

    b. Move to the **Page** tab and select the **Continue From Previous Page in Book** radio button.

7. Click **Set.**

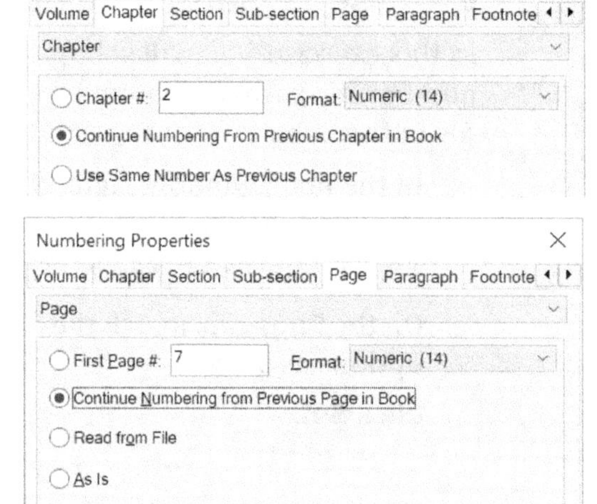

Exercise 15: Update the Book

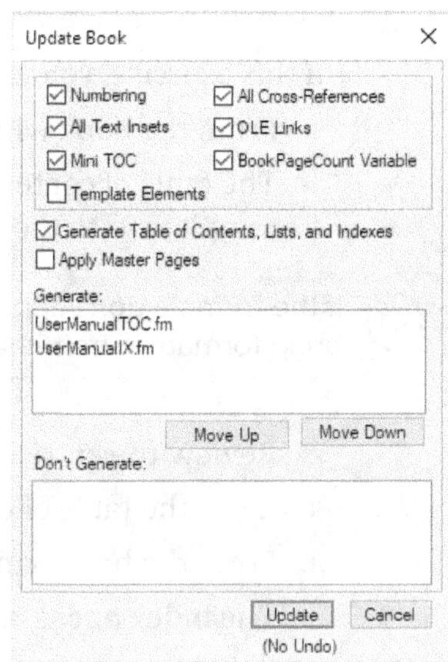

In this exercise, you will update the book to fix page sides and numbering and update paragraph numbering and populate the generated files.

1. Select the **Update Book** button ( ) in the book file.

   The **Update Book** dialog appears.

2. Click **Update.**

   Messages appear at the bottom of the book window showing the progress of the update.

   When done, the **Structure View** continues to display each file as **Chapter**, the highest-level element in each file, except for the generated files which still display as **BOOK-COMPONENT.**

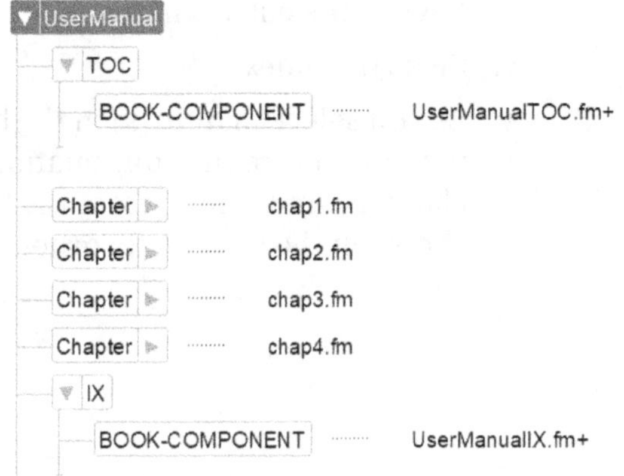

3. With the book window active, shift-click the **File** menu, choose **Save All Files in Book.**

If you get a book error log related to inconsistent numbering properties, select the "Skip Remaining" checkbox and continue. Once you save all files in book, the message will be resolved.

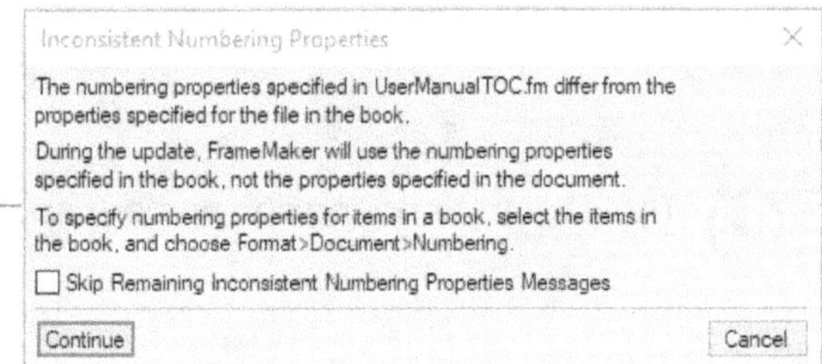

Exercise 16: Reviewing the Book Files

In this exercise, you will open and check the newly generated table of contents and index.

1. From the book window, double-click the table of contents file.

   The table of contents opens, - with formatting applied - because you saved the table of contents template with the correct file name.

If the formats don't appear correctly for either the TOC or IX, you can use **File > Import > Formats** to bring formats in from the corresponding template file.

2. Review the entries in the table of contents.
3. Close the table of contents.
4. From the book window, double-click the index file.

   The index opens, already formatted because you saved the index template with the correct file name.
5. Review the entries in the index.
6. Close the index.
7. As you select chapter(s) in the book window, notice the continuation of the chapter numbering across the files shown in the lower left corner.

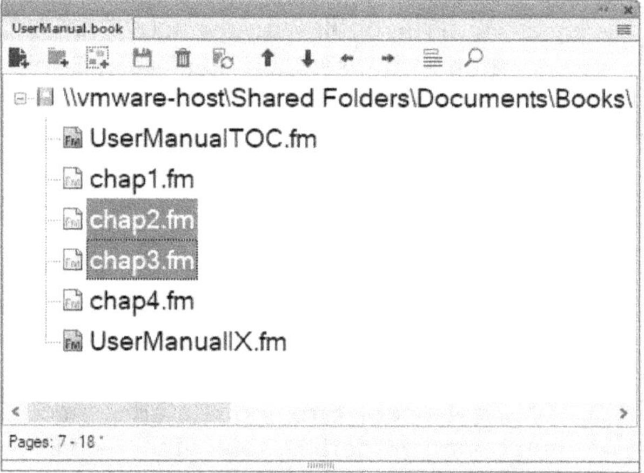

The page numbering shown corresponds to the page numbering set/shown in each individual chapter.

Chapter 15: Structuring Unstructured Data

Introduction

In this chapter, you will set up a conversion table and define object and element mapping in it. After defining the conversion table, you will use its mapping to add structure to unstructured documents.

Objectives

- Learn conversion rule syntax
- Generate a conversion table
- Learn how to create a conversion table from scratch
- Structure a currently unstructured document
- Structure a group of unstructured files
- Structure an unstructured book

Overview

There are two ways to wrap unstructured data directly in FrameMaker:

- Method 1—Manually, element by element. This method is only good for very small conversions.
- Method 2—Automatically, using the tools available in **Structure > Utilities**

Automatic wrapping requires a conversion table, which is usually created by an application developer.

A conversion table:

- Provides mappings to automate the task of adding structure to unstructured documents
- Uses paragraph and character tags, and object types (such as equations or footnotes), to identify how to wrap document components in elements
- Uses context to wrap child elements in parent elements

Here is the organization of the conversion table itself:

- A regular table, with at least 3 columns and 1 body row
- It has additional columns and heading/footing rows for comments and labeling
- Each body row holds 1 rule. Here is a breakdown of table organization:

Column 1	Column 2	Column 3
specifies document object, child element, or sequence to wrap	specifies element in which to wrap	specifies optional qualifier ("nickname") to use as temporary label

A conversion table documument can be split up into several tables with text or graphics in between for comments

- It cannot have any tables other than conversion tables
- The table must be saved before it can be used

Rule Syntax—Character Restrictions

- Tags are case-sensitive and tags in tables must exactly match tags in docs to be converted.
- Qualifier tags are case-sensitive and two occurrences of one qualifier must match exactly

Qualifiers for these special characters in tags: ( ) & | , * + ? % [ ] : \

- Are allowed in format tags and qualifier tags— but only if preceded by a backslash (\) in the table
- Are not allowed in element tags

A space character in tags does not need to be preceded with a backslash
(For example, the tag **Format A** is valid)

Wildcard character (%) in Tags:

- Use % as in format or element tag to match zero, one, or more characters (similar to * in general rule)
- For example, P:%Body matches paragraphs with the format tag Body, FirstBody, or BulletBody

Methods for Producing a Conversion Table

- Method 1—FrameMaker Generates Initial Table (easiest, recommended, and is the method used in here). Once created, it can be modified as needed by hand.
- Method 2—Create a conversion table from scratch (requires creating and populating entries for all catalog entries, and not recommended)

Exercise 1: Generating Initial Conversion Table

In this exercise, you will generate a conversion table from an unstructured document.

1. From your class files directory, open **wraptest.fm**.
   a. From the **File** menu, choose **Open**.
      The **Open** dialog appears.
   b. If necessary, change to your class files directory.
   c. Double-click **wraptest.fm**.

For instructions on downloading class files, see "Downloading class files" on page 1.

The document appears.

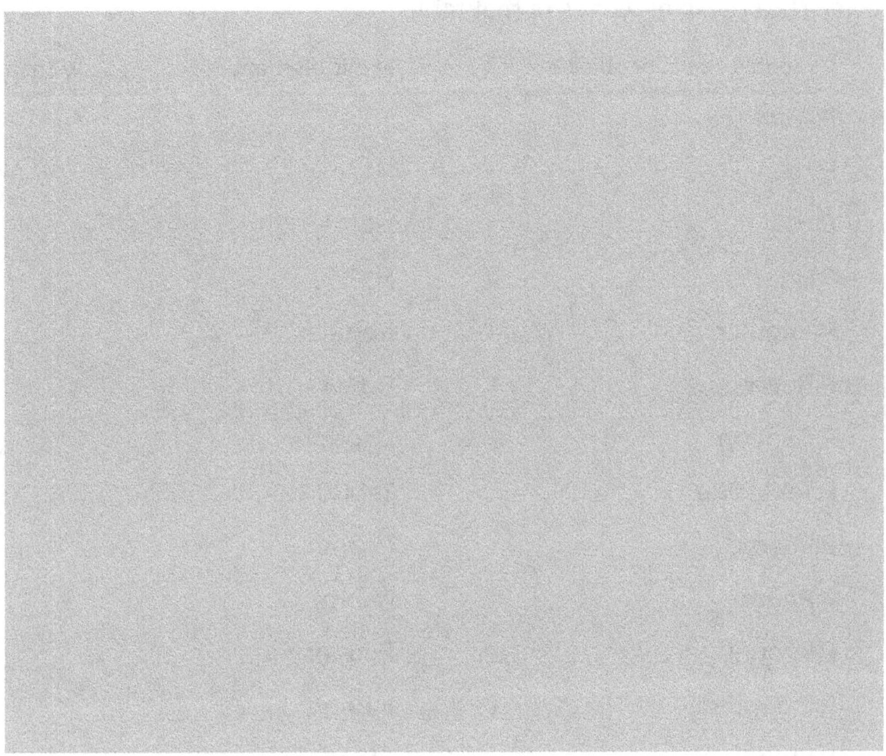

2. Open the **Structure View** and **Elements** panels.

> Both the **Structure View** and **Elements** panels are blank because this is an unstructured document.

3. Scroll through the document, clicking in various paragraphs to identify the paragraph formats being used.
4. From the **Structure** menu, choose **Utilities>Generate Conversion Table.**

   The **Generate Conversion Table** dialog appears.
5. Select the **Generate New Conversion Table** option and click **Generate.**

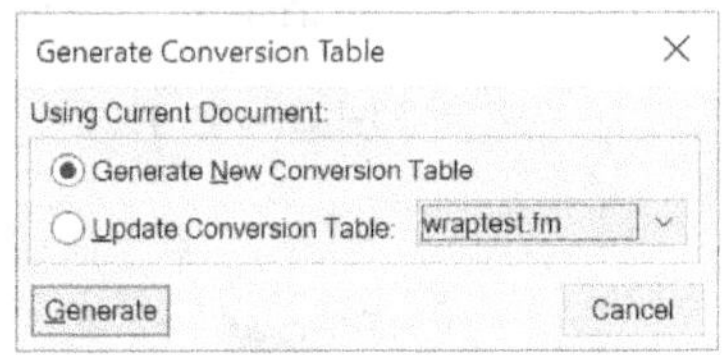

A conversion table, based on your document's format ta-gs, appears.

Here is a recap of what's in that table.

Wrap this object or objects	In this element	With this qualifier
P:Title	Title	
P:H1	H1	
P:H2	H2	
P:H3	H3	
P:Regular	Regular	
P:Bulleted	Bulleted	
P:Caption	Caption	
P:TableTitle	TableTitle	
P:Pname	Pname	
P:Pnum	Pnum	
P:Pcount	Pcount	
P:Indented	Indented	
P:StepFirst	StepFirst	
P:Step	Step	
P:Note	Note	
C:Emphasis	Emphasis	
C:WarnNote	WarnNote	
X:ElemNumTextPage	ElemNumTextPage	
M:Index	Index	
M:Cross-Ref	Cross-Ref	
G:	GRAPHIC	
F:flow	FOOTNOTE	
T:Format A	FormatA	
TT:	TITLE	
TH:	HEADING	
TB:	BODY	
TF:	FOOTING	

Wrap this object or objects	In this element	With this qualifier
TR:	ROW	
TC:	CELL	

6. Save this new untitled document into your class files directory with the new filename `Conversion Table.fm`
   a. From the **File** menu, choose **Save As.**

      The **Save Document** dialog appears.
   b. If necessary, change to your class files directory.
   c. In the **Save in File** field , delete the current file name and type: `Conversion Table.fm`
   d. Click **Save.**

Typing the Conversion Rules

Conventions for Column 1 of the conversion table:

- Type a one- or two-letter code to identify the type of item
- Type a format (optional) to narrow the definition

For this object	Use this Code	Followed by optional
Paragraph	P:	Paragraph format tag
Text range	C:	Character format tag
Table	T:	Table format tag
Table title	TT:	(none)
Table heading	TH:	(none)
Table body	TB:	(none)
Table row	TR:	(none)
Table cell	TC:	(none)
System variable	SV:	Variable format name
User variable	UV:	Variable format name
Graphic (anchored frame or imported object)	G:	(none)
Footnote	F:	Location of footnote: Table or Flow
Marker	M:	Marker type
Cross-reference	X:	Cross-reference format
Text Inset	TI:	(none)

For this object	Use this Code	Followed by optional
Equation	Q:	Size of equation: Small, Medium, or Large
Element	E:	Element tag (used to wrap elements in higher-level elements)

Conventions for Column 2 of the conversion table:

- Type the object identifier E: (optional)
- Followed by an element tag

Conventions for Column 3 of the conversion table:

- Type a qualifier (optional) for the new element tag
- Qualifiers are used in later rules to differentiate elements of the same name when wrapping in higher-level elements

Conventions for wrapping a sequence of elements, in Column 1:

- Type E: for element
- Type an element tag
- Type a qualifier (optional) in brackets
- Add more element tags with code identifiers
- Use these symbols to further describe the sequence

Symbol	Meaning
Plus sign (+)	Item is required and can occur more than once
Question mark (?)	Item is optional and can occur once
Asterisk (*)	Item is optional and can occur more than once
Comma (,)	Items must occur in the order given
Ampersand (&)	Items can occur in any order
Vertical bar (|)	Any one of the items in the sequence can occur
Parentheses	Beginning and end of a sequence

To specify an attribute for an element, in Column 2:

- Type the attribute name and value in brackets after the element tag in the second column of the table
- Separate the name and value with an equal sign, and enclose the value in double quotation marks

If naming a table element from one or more child elements, in Column 1:

- Type the object identifier T:
- Followed by E:

- Followed by an element tag
- Type a qualifier (optional) in brackets

To promote anchored objects, in Column 2:

- Type the element tag for the table or graphic
- Add the keyword "**promote**" in parentheses after the element tag for the table or graphic

To flag format overrides, in Column 1:

- Add the rule "**flag paragraph format overrides**"
- Add the rule "**flag character format overrides**"

To wrap untagged text:

- In Column 1, add the rule "**untagged character formatting**"
- In Column 2, add an element tag

Exercise 2: Writing Rules for Text Ranges

In this exercise, you will modify the existing rules for text ranges to match the element tags in **Final EDD.fm**.

1. Locate the following rule:

Wrap this object or objects	In this element	With this qualifier
C:Emphasis	Emphasis	

2. Text using the character tag **Emphasis** should be wrapped in a **Term**. Once wrapped, the **Emphasis** character tag is not necessary, as the **TextFormatRules** for **Term** specify the italic formatting.

   For this reason, rewrite the rule above as follows:

C:Emphasis	Term	

3. Text using character tag **WarnNote** does not need to be wrapped. This text corresponds with the prefix for the **Term** element. This rule is no longer needed, as the **PrefixRules** for **Term** specify the red color for the prefix.

   For this reason, find and delete this following rule entirely:

Wrap this object or objects	In this element	With this qualifier
C:WarnNote	Term	

4. Save your changes.

Exercise 3: Writing Rules for Paragraphs

In this exercise, you will modify the existing rules for paragraphs to match the element tags in **Final EDD.fm**.

1. Locate the following rules:

Wrap this object or objects	In this element	With this qualifier
P:H1	H1	
P:H2	H2	
P:H3	H3	

Text using paragraph tag **H1** should be wrapped in **Head**. To differentiate this **Head** from **Head** elements at other nesting levels, give it a qualifier of **SH1**.

Text using paragraph tag **H2** should be wrapped in **Head**. To differentiate this **Head** from other **Head** elements, give it a qualifier of **SH2**.

Text using paragraph tag **H3** should be wrapped in **Head**. To differentiate this **Head** from other **Head** elements, give it a qualifier of **SH3**.

For these reasons, rewrite these rules as follows:

P:H1	Head	SH1
P:H2	Head	SH2
P:H3	Head	SH3

2. Locate the following rules:

Wrap this object or objects	In this element	With this qualifier
P:Regular	Regular	
P:Bulleted	Bulleted	
P:Indented	Indented	
P:StepFirst	StepFirst	
P:Step	Step	

   Text using paragraph tag **Regular** should be wrapped in **Para**. To differentiate this **Para** from **Para** elements within lists, give it a qualifier of **Regular**.

   Text using paragraph tag **Bulleted** should be wrapped in **Para**. To differentiate this **Para** which appears within a bulleted list, give it a qualifier of **Bull**.

   Text using paragraph tag **Indented** should be wrapped in **Para**. To differentiate this **Para** which serves as a substep to bulleted or numbered items in **List** elements, give it a qualifier of **Substep**.

   Text using paragraph tag **StepFirst** should be wrapped in **Para**. To differentiate this **Para** which restarts a numbered list, give it a qualifier of **Step1**.

   Text using paragraph tag **Step** should be wrapped in **Para**. To differentiate this **Para** which continues a numbered list, give it a qualifier of **Step**.

   For these reasons, rewrite these rules as follows:

Wrap this object or objects	In this element	With this qualifier
P:Regular	Para	Regular
P:Bulleted	Para	Bull
P:Indented	Para	Substep
P:StepFirst	Para	Step1
P:Step	Para	Step

3. Find the following rule:

Wrap this object or objects	In this element	With this qualifier
P:Note	Note	

   Text using paragraph tag **Note** should be wrapped in **Term**.

   For this reason, rewrite this rule as follows:

Wrap this object or objects	In this element	With this qualifier
P:Note	Term	

4. Save your changes.

Exercise 4: Writing Rules for Footnotes

In this exercise, you will modify the existing rules for footnotes to match the element tags in `Conversion Table.fm`.

1. Locate the following rule:

Wrap this object or objects	In this element	With this qualifier
F:flow	FOOTNOTE	

Footnotes in the main flow should be wrapped in **Footnote** (initial capital letter only).

For this reason, rewrite this rule as follows:

F:flow	Footnote	

Additionally, although you did not have table footnotes in this particular example, you may in other unstructured documents for which you want to use the same conversion table.

For this reason, add the following rule:

F:table	Footnote	

2. Save your changes.

Exercise 5: Writing Rules for Cross-References

In this exercise, you will modify the existing rules for cross-references to match the element tags in **Final EDD.fm**.

1. Locate the following rule:

Wrap this object or objects	In this element	With this qualifier
X:ElemNumTextPage	ElemNumTextPage	

Cross-references need to be wrapped in **XRef**, so rewrite the rule as follows:

X:ElemNumTextPage	XRef	

2. Save your changes.

Exercise 6: Writing Rules for Equations

In this exercise, you will create a rule for equations to match the element tags in **Final EDD.fm**.

1. Add the following rule anywhere in the table (creating a new table row, if necessary):

Wrap this object or objects	In this element	With this qualifier
Q:	EQ (promote)	

You could specify an equation size of **Small**, **Medium** or **Large**, but you want to wrap all equations in **EQ**, regardless of their size.

Adding the word "**(promote)**" makes **EQ** elements a sibling, rather than a child, of the paragraph element in which they are anchored.

2. Save your changes.

Exercise 7: Writing Rules for Graphics

In this exercise, you will modify the existing rule for graphics to match the element tags in **Final EDD.fm.**

1. Locate the following rule:

Wrap this object or objects	In this element	With this qualifier
G:	GRAPHIC	

Graphics should be wrapped in Graphic (initial capital letter only). Adding the word "**(promote)**" makes them a sibling, rather than a child, of the paragraph element in which they are anchored. According to the EDD, they should be a sibling of **Caption**, not a child.

For this reason, rewrite this rule as follows:

G:	Graphic (promote)	

2. Save your changes.

Exercise 8: Writing Rules for Markers

In this exercise, you will modify the existing rule for index markers and delete the rule for cross-reference markers to match the element tags in **Final EDD.fm.**

1. Locate the following rules:

Wrap this object or objects	In this element	With this qualifier
M:Index	Index	
M:Cross-Ref	Cross-Ref	

Cross-Ref markers should not be wrapped. You will be using element-based cross-referencing based on attributes, so delete the second rule entirely.

2. Index markers should be wrapped in an **IndexEntry** element, so rewrite the Index rule as follows:

M:Index	IndexEntry	

3. Save your changes.

Exercise 9: Writing Rules for Tables and Table Parts

In this exercise, you will modify the existing rules for tables and table parts to match the element tags in **Final EDD.fm**.

1. Locate the following rules:

Wrap this object or objects	In this element	With this qualifier
P:TableTitle	TableTitle	
P:Pname	Pname	
P:Pnum	Pnum	
P:Pcount	Pcount	
T:Format A	FormatA	

Because you are not allowing elements within the TableTitle, just <TEXT>, you do not need a rule for wrapping the P:TableTitle in an element.

Because you are not allowing elements within the cells of the table, just <TEXT>, you do not need a rule for wrapping the cells' paragraphs in elements before wrapping them in table cell elements.

For these reasons, delete the following rules:

P:TableTitle	TableTitle
P:Pname	Pname
P:Pnum	Pnum
P:Pcount	Pcount

2. Table elements should be wrapped in a **Table** element. Adding the word "(promote)" makes them a sibling, rather than a child, of the paragraph element in which they are anchored.

For this reason, rewrite the Table rule as follows:

T:Format A	Table (promote)

3. Locate the following rules:

Wrap this object or objects	In this element	With this qualifier
TT:	TITLE	
TH:	HEADING	
TB:	BODY	
TF:	FOOTING	

Wrap this object or objects	In this element	With this qualifier
TR:	ROW	
TC:	CELL	

These generic elements are already defined specifically in your EDD, so rewrite the rules for the table heading, table body, and table footing elements as follows:

TH:	TableHeading	
TB:	TableBody	
TF:	TableFooting	

4. Rewrite the table row rule as follows to match the EDD:

TR:	TableRow	

5. The content (using the Pname format) in the cell will be wrapped directly in the table cell element with the tag **Name**, so rewrite the table cell rule as follows to match the EDD:

TC:P:Pname	Name	

6. The paragraph (using Pnum format) in the cell will be wrapped directly in the table cell element with the element tag **Num**. Pcount will be wrapped in **Count**, so add the following two rules:

TC:P:Pnum	Num	
TC:P:Pcount	Count	

7. Save your changes.

Exercise 10: Wrapping Captions and Graphics in Figures

In this exercise, you will add a rule to group the **Caption** and **Graphic** together in a **Figure** element.

Although your EDD requires the **Caption**, you will make it optional in the wrapping process. In case an author forgot the **Caption**, **Graphic** will still be wrapped in **Figure**, with a square hole displaying in the **Structure View** where the **Caption** is missing.

Your goal is to wrap as much as possible, even if the unstructured document does not quite conform to the element definitions defined in the EDD.

In the end, you will have less manual wrapping to do.

1. Add the following rule:

Wrap this object or objects	In this element	With this qualifier
E:Caption?, E:Graphic	Figure	

2. Save your changes.

Note this is the first instance of wrapping multiple child elements in a parent element. This is what allows you to create sophisticated patterns that match your content model.

Exercise 11: Wrapping Paras in Items in Lists

In this exercise, you will add several rules to wrap the various **Para** elements (by qualifier) in **Item**. Then, you will wrap the **Item** in a **List**.

1. Create rules for the following content that wraps in **Item**:

   The first rule wraps a **Para** with qualifier **Step1**—followed by zero or more of **Para** with qualifier **Substep**, **Term**, and **Figure** elements—in an **Item** with qualifier **First**.

   The second rule wraps a **Para** with qualifier **Step**—followed by zero or more of **Para** with qualifier **Substep**, **Term**, and **Figure** elements—in an **Item** with qualifier **Additional.**

   The third rule wraps a **Para** with qualifier **Bull**—followed by zero or more of **Para** with qualifier **Substep**, **Term**, and **Figure** elements—in an **Item** with qualifier **BullItem** (watch the spelling).

   Add the following three rules:

Wrap this object or objects	In this element	With this qualifier
E:Para[Step1], (E:Para[Substep] | E:Term | E:Figure)*	Item	First
E:Para[Step], (E:Para[Substep] | E:Term | E:Figure)*	Item	Additional
E:Para[Bull], (E:Para[Substep] | E:Term | E:Figure)*	Item	BullItem

2. Create rules for the following content that wraps in **List**:

   The first rule wraps one or more **Item** elements with qualifier **BullItem** in a **List** with the **ListType** attribute and a value of **Bulleted**, which drives the formatting of bulleted lists in your EDD.

   The second rule wraps one **Item** element with qualifier **First**—followed by zero or more **Item** elements with qualifier **Additional**—in a **List** with the **ListType** attribute and a value of **Numbered**, which drives the formatting of numbered lists in your EDD.

   Again, to maximize the amount of automatic wrapping, the conversion rules for the number of items in a list can be less restrictive than the actual rules in the EDD. (EDD requires two, conversion rules require only one.) Missing required elements will display as square holes in the **Structure View**.

   Add the following two rules:

Wrap this object or objects	In this element	With this qualifier
E:Item[BullItem]+	List [ListType = "Bulleted"]	
E:Item[First], E:Item[Additional]*	List [ListType = "Numbered"]	

3. Save your changes.

Exercise 12: Wrapping Heads and Their Siblings in Sections

In this exercise, you will add several rules to wrap the various **Head** elements (by qualifier) and their siblings in a **Section**.

Create rules for the following content that wraps in **Section**:

The first rule wraps a 3rd-level **Head** and its siblings in a **Section** with qualifier **Section3**.

The second rule wraps a 2nd-level **Head** and its siblings (some of which can be a 3rd-level **Section**) in a **Section** with qualifier **Section2**.

The third rule wraps a 1st-level **Head** and its siblings (some of which can be a 2nd-level **Section**) in a **Section** with qualifier **Section1**.

Again, to maximize the amount of automatic wrapping, the conversion rules for the number of a subordinate-level **Section** within a parent **Section** are less restrictive than those in the EDD.

Add the following three rules:

Wrap this object or objects	In this element	With this qualifier
E:Head[SH3], (E:Para[Regular] | E:List | E:Table | E:Figure | E:EQ | E:Term)+	Section	Section3
E:Head[SH2], (E:Para[Regular] | E:List | E:Table | E:Figure | E:EQ | E:Term | Section[Section3])+	Section	Section2
E:Head[SH1], (E:Para[Regular] | E:List | E:Table | E:Figure | E:EQ | E:Term | Section[Section2])+	Section	Section1

1. Save your changes.

Exercise 13: Wrapping the Highest-Level Element

In this exercise, you will add a final rule to wrap up the entire document in a **Chapter**.

To maximize the amount of automatic wrapping, the conversion rule doesn't require more than one **Section** in the **Chapter**. As long as you import element definitions from a structured template or EDD in the document you are wrapping, missing and misplaced elements will be found when validating.

1. Add the following rule:

E:Title, E:Section[Section1]+	Chapter

2. Save your changes.

Completed Conversion Table

Wrap this object or objects	In this element	With this qualifier
P:Title	Title	
P:H1	Head	SH1
P:H2	Head	SH2
P:H3	Head	SH3
P:Regular	Para	Regular
P:Bulleted	Para	Bull
P:Caption	Caption	
P:Indented	Para	Substep
P:StepFirst	Para	Step1
P:Step	Para	Step
P:Note	Term	
C:Emphasis	Term	
X:ElemNumTextPage	XRef	
M:Index	IndexEntry	
Q:	EQ	
G:	Graphic (promote)	
F:flow	Footnote	
F:table	Footnote	
T:Format A	Table (promote)	
TT:	TableTitle	
TH:	TableHeading	
TB:	TableBody	
TF:	TableFooting	
TR:	TableRow	
TC:P:Pname	Name	
TC:P:Pnum	PartNum	
TC:P:Pcount	Count	
E:Caption, E:Graphic	Figure	

Wrap this object or objects	In this element	With this qualifier
E:Para[Step1], (E:Para[Substep] | E:Term | E:Figure)*	Item	First
E:Para[Step], (E:Para[Substep] | E:Term | E:Figure)*	Item	Additional
E:Para[Bull], (E:Para[Substep] | E:Term | E:Figure)*	Item	BullItem
E:Item[BullItem]+	List [ListType = "Bulleted"]	
E:Item[First], E:Item[Additional]*	List [ListType = "Numbered"]	
E:Head[SH3], (E:Para[Regular] | E:List | E:Table | E:Figure | E:EQ | E:Term)+	Section	Section3
E:Head[SH2], (E:Para[Regular] | E:List | E:Table | E:Figure | E:EQ | E:Term | Section[Section3])+	Section	Section2
E:Head[SH1], (E:Para[Regular] | E:List | E:Table | E:Figure | E:EQ | E:Term | Section[Section2])+	Section	Section1
E:Title, E:Section[Section1]+	Chapter	

Structuring Unstructured Documents

With a conversion table, you can structure:

- Single files
- Groups of files
- Books and their component files

Exercise 14: Structuring Current Unstructured Document

In this exercise, you will structure a single file using your conversion table.

1. If not still open, from your class files directory, open **Conversion Table.fm.**

   If you did not finish the conversion table, please open **Final Conversion Table.fm Conversion Table.fm** instead, and save it in your class files directory as **Conversion Table.fm.**

If you have difficulty with the table you created, the **Final Conversion Table.fm** document can help you troubleshoot your conversion rules to find typos or mistakes.

2. From your class files directory, open **Final EDD.fm.**

Final EDD.fm has slight structural differences from the EDD you finished up developing in the last chapter. Use **Final EDD.fm** and not **EDD.fm** for best results.

3. If not still open, from your class files directory, open **wraptest.fm.**

4. In **wraptest.fm**, import element definitions from **Final EDD.fm**

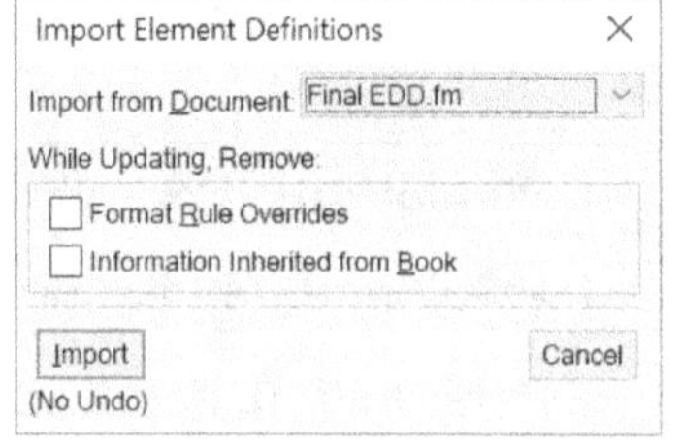

   a. From the **File** menu in `wraptest.fm`, choose **Import > Element Definitions.**
   b. From the **Import from Document** popup menu, choose **Final EDD.fm.**
   c. Click **Import.**

      An alert box appears indicating "**Element definitions have been imported from the EDD**"
   d. Click **OK** to close any alerts.
5. Save your changes.

By importing your EDD into the document to be structured, you are passing the EDD into your converted document. This may be significant in your own work, as you'll likely have many iterations of testing, and importing the EDD into each resulting converted document can become tedious.

6. From the **Structure** menu in **wraptest.fm**, choose **Utilities > Structure Current Document.**

   The **Structure Current Document** dialog appears
7. From the **Conversion Table Document** popup menu, choose **Conversion Table.fm.** (Choose Final Conversion Table.fm, if you didn't earlier complete your own conversion table.)

Structure Current Document
Conversion Table Document: Conversion.fm
Add Structure
Cancel

8. Click **Add Structure.**

   An alert appears indicating "**Operation completed normally.**"
9. Click **OK** to dismiss the alert.

   If there were errors, a log file appears diagnosing the errors.
10. If necessary, modify **Conversion Table.fm** and repeat the conversion.

    A new **NoName** document appears with the initial file's contents wrapped according to the rules in the conversion table.
11. Save the file as `wraptest.out.fm`.
12. Validate and correct any validation errors.

    You will need to correct a missing attribute value for the **Author** attribute and fix some unresolved cross-references, relinking them to **Table** and **Figure** elements, rather than paragraphs.
13. If elements do not appear to be wrapped correctly, modify **Conversion Table.fm** and restructure the **wraptest.fm** file, not the output file which is already wrapped.
14. Save and close all your files.

www.ingramcontent.com/pod-product-compliance
Lightning Source LLC
Chambersburg PA
CBHW082244060726
47598CB00016B/2773
* 9 7 8 1 9 5 3 4 8 8 0 3 9 *